T0335825

Static and Dynamic Problems of

NANOBEAMS and NANOPLATES

Static and Dynamic Problems of
NANOBEAMS and NANOPLATES

Snehashish Chakraverty
Laxmi Behera
National Institute of Technology Rourkela, India

World Scientific

NEW JERSEY • LONDON • SINGAPORE • BEIJING • SHANGHAI • HONG KONG • TAIPEI • CHENNAI • TOKYO

Published by

World Scientific Publishing Co. Pte. Ltd.
5 Toh Tuck Link, Singapore 596224
USA office: 27 Warren Street, Suite 401-402, Hackensack, NJ 07601
UK office: 57 Shelton Street, Covent Garden, London WC2H 9HE

Library of Congress Cataloging-in-Publication Data
Names: Chakraverty, Snehashish, author. | Behera, Laxmi, author.
Title: Static and dynamic problems of nanobeams and nanoplates / Snehashish Chakraverty (National Institute of Technology Rourkela, India), Laxmi Behera (National Institute of Technology Rourkela, India).
Description: New Jersey : World Scientific, [2016] | Includes bibliographical references.
Identifiers: LCCN 2016029080 | ISBN 9789813143913 (hc : alk. paper)
Subjects: LCSH: Nanostructured materials--Mechanical properties. | Elastic analysis (Engineering) | Micromechanics. | Elastic plates and shells. | Elastic rods and wires.
Classification: LCC TA418.9.N35 C369 2016 | DDC 620.1/153--dc23
LC record available at https://lccn.loc.gov/2016029080

British Library Cataloguing-in-Publication Data
A catalogue record for this book is available from the British Library.

Desk Editors: Suraj Kumar/Rhaimie Wahap

Typeset by Stallion Press
Email: enquiries@stallionpress.com

Printed in Singapore

Preface

In recent decades, nanotechnology is becoming a challenging area of research. Structural elements such as beams and plates in micro or nanoscale have a vast range of practical applications. In general, conducting experiments at nanoscale size is difficult and so development of related mathematical models is helpful. Among various size dependent theories, nonlocal elasticity theory pioneered by Eringen is being increasingly used for reliable and better analysis of nanostructures. Finding solutions of governing partial differential equations are the key factor in static and dynamic analyses of nanostructures. In this regard, exact or closed-form solutions for these differential equations are sometimes not possible. As such, approximate methods have been developed by various researchers. But, the methods may not always handle all sets of boundary conditions. Accordingly, computationally efficient numerical methods have been developed recently for better understanding of static and dynamic behaviors of nanostructures. The present book addresses bending, buckling and vibration of nanobeams and nanoplates by solving the corresponding partial differential equations. In the above regard, various beam and plate theories are discussed for the analysis and corresponding results are reported. Complicating effects such as non-uniform material properties are considered in some of the problems. Other complicating effects such as surrounding medium and temperature are important in the nanotechnology applications. Accordingly, the effect of these complicating effects has also been investigated in detail.

Rayleigh–Ritz and differential quadrature methods have been discussed here in particular to solve the above said problems. In the Rayleigh–Ritz method, simple and boundary characteristic orthogonal polynomials have been used as shape functions. Boundary characteristic orthogonal polynomials in the Rayleigh–Ritz method has some advantages over other shape functions. On the other hand, Differential Quadrature (DQ) method is a computationally efficient method which can be used to solve higher order partial differential equations that may handle all sets of classical boundary conditions. Accordingly, DQ method has also been reported to be of use in solving few problems of nanobeams.

In view of the above, systematic study of bending, buckling and vibration of nanobeams and nanoplates are included here. Accordingly, this book consists of 10 chapters. Recently, effort has been made by various researchers to solve these types of problems but a lot of important information is still missing in the existing books. Further, some of the known methods are computationally expensive and may not handle all sets of classical boundary conditions. The purpose of the present work is to fill these gaps. Accordingly, the contents of the 10 chapters are summarized below:

Chapter 1 includes a brief introduction of nanostructures and their applications. This chapter highlights advantage of small scale effect on the nanotechnology applications and importance of nonlocal elasticity theory. Some of the preliminaries related to various beam and plate theories have been discussed. Chapter 2 presents few analytical methods developed by previous researchers to handle some of the beam and plate theories. Numerical methods such as Rayleigh–Ritz and differential quadrature are discussed in Chapter 3. Advantage of these numerical methods and systematic procedure for applying boundary conditions has also been highlighted.

Chapter 4 addresses bending of nanobeams. Bending analysis has been carried out based on Euler–Bernoulli and Timoshenko nonlocal beam theories by using boundary characteristic orthogonal polynomials as shape functions in the Rayleigh–Ritz method. Buckling of nanobeams has been discussed in Chapter 5. Differential quadrature method has been employed to study buckling analysis of non-uniform nanobeams based on four beam theories such as Euler–Bernoulli, Timoshenko, Reddy and Levinson. Buckling analysis of embedded nanobeams under the influence of temperature has also been investigated based on Euler–Bernoulli, Timoshenko and Reddy beam theories. In these problems, boundary characteristic and Chebyshev polynomials have been used in the Rayleigh–Ritz method respectively for Euler–Bernoulli and Timoshenko beam theories. Further, differential quadrature method has been employed for buckling analysis of nanobeams embedded in elastic foundations based on nonlocal Reddy beam theory. Effects of temperature and foundation parameters on the buckling load parameter have also been addressed.

In Chapter 6, vibration of nanobeams has been investigated. Rayleigh–Ritz method with simple polynomials and boundary characteristic orthogonal polynomials has been implemented to compute vibration characteristics of Euler–Bernoulli and Timoshenko nanobeams. Differential quadrature method has been used to investigate four types of nonlocal beam theories such as Euler–Bernoulli, Timoshenko, Reddy and Levinson. In these problems, differential equations are converted into single unknown variable and boundary conditions have been substituted in the coefficient matrices. Vibration of nanobeams with complicating effects is included in Chapter 7. The complicating effects such as non-uniformity, temperature and Winker as well as Pasternak foundations have been taken into consideration. At first, Rayleigh–Ritz method with boundary characteristic orthogonal polynomials has been used to investigate vibration of non-uniform Euler–Bernoulli and Timoshenko nanobeams. Non-uniformity is assumed to arise due to linear and quadratic variations in Young's modulus and density with space coordinate. Next, vibration analysis of embedded nanobeams with elastic foundations has been analyzed in the influence of temperature. Rayleigh–Ritz method has been used with boundary characteristic orthogonal polynomials and Chebyshev polynomials as shape functions for Euler–Bernoulli and Timoshenko nonlocal beam theories respectively. On the other hand, differential quadrature method has been implemented for vibration analysis of embedded nanobeams with elastic foundation based on Reddy nonlocal beam theory. Effects of temperature, Winkler and Pasternak coefficients on the frequency parameter have also been addressed.

Chapter 8 deals with bending and buckling problems of nanoplates based on classical plate theory in conjunction with nonlocal elasticity theory of Eringen. In this chapter, rectangular nanoplates are considered. Two-dimensional simple polynomials have been implemented in the Rayleigh–Ritz method to investigate these problems. Effects of length, nonlocal parameter and aspect ratio on the non-dimensional maximum deflection have been discussed. Similarly, effects of length, nonlocal parameter, aspect ratio, stiffness ratio and foundation parameters on the buckling loads have also been included.

Vibration of rectangular nanoplate has been studied in Chapter 9 based on classical plate theory in conjunction with nonlocal elasticity theory of Eringen. Here, two-dimensional simple polynomials have been used as shape functions in the Rayleigh–Ritz method. Finally, Chapter 10 includes vibration of nanoplates with various complicating effects. Two dimensional simple polynomials and boundary characteristic orthogonal polynomials have been used as shape functions in the Rayleigh–Ritz method. Complicating effects such as non-uniformity taking linear and quadratic variations of Young's modulus and density along space coordinate are addressed. Investigation has also been done for the nanoplate when it is embedded in elastic foundation such as Winkler and Pasternak.

This book aims to provide a new direction for the readers with basic concepts of nanostructures and related equations, solutions and their applications in a systematic manner along with the recent developments. This book may be an essential text and reference for students, scholars, practitioners, researchers and academicians in the assorted fields of engineering and science. The book provides comprehensive results, up to date and self-contained review of the topic along with application oriented treatment of nanobeams and nanoplates to be used in various domains.

S. Chakraverty
Laxmi Behera
September, 2016

About the Authors

Snehashish Chakraverty, Ph.D., is Professor in the Department of Mathematics, National Institute of Technology Rourkela in India. He did his Ph.D. from IIT Roorkee and did post-doctoral research from ISVR, University of Southampton, U.K. and Concordia University, Canada. He was visiting Professor at Concordia, McGill and Johannesburg universities and also was with Central Building Research Institute, Roorkee. He has published six books, 240 research papers, and has been the reviewer of many international journals. Prof. Chakraverty has been the recipient of CSIR Young Scientist, BOYSCAST, UCOST, Golden Jubilee CBRI, INSA International Bilateral Exchange, Platinum Jubilee ISCA Lecture and Roorkee University gold medal awards. He is the Chief Editor of International Journal of Fuzzy Computation and Modelling (IJFCM), Inderscience Publisher, Switzerland (http://www.inderscience.com/ijfcm) and happens to be the Guest Editor for few other journals. He was the President of the Section of Mathematical Sciences (including Statistics) of Indian Science Congress (2015–2016) and was the Vice President — Orissa Mathematical Society (2011–2013). He has already guided 11 Ph.D. students and 9 are ongoing. Prof. Chakraverty has undertaken around 17 research projects as Principal Investigator funded by international and national agencies totalling about Rs 1.5 crores. His research areas include differential equations, numerical analysis, soft computing, vibration and inverse problems.

Laxmi Behera completed her Ph.D. from National Institute of Technology Rourkela, Odisha, India in 2016. Prior to this, she has qualified in Graduate Aptitude test in Engineering (GATE-2011). After completing graduation from Regional Institute of Education (Bhubaneswar), her career started from National Institute of Technology Rourkela and she did M.Sc. in Mathematics. She has published nine research papers (till date) in international refereed journals. Her present research areas include numerical analysis, differential equations, mathematical modeling, vibration and nanotechnology.

Contents

Chapter 1

Introduction

1.1 Preliminaries of Nanomaterials and Structural Members

Nanotechnology is concerned with the fabrication of functional materials and systems at the atomic and molecular levels. Recently, development of nanotechnology enables a new generation of materials with revolutionary properties and devices with enhanced functionality (Ansari and Sahmani 2011) having a vast range of applications, such as in medicine, electronics, biomaterials, and energy production (Şimşek and Yurtcu 2013). Recently, nanomaterials have encouraged the interest of the scientific researchers in mathematics, physics, chemistry, and engineering. These nanomaterials have outstanding mechanical, chemical, electrical, optical, and electronic properties. Owing to their properties, the nanomaterials are perceived to be the components for various nanoelectromechanical systems and nanocomposites. Some of the common examples of these nanomaterials are nanoparticles, nanowires, nanotubes (viz. carbon nanotubes (CNTs), ZnO nanotubes), etc. Nanomaterials are the basis materials of many nanoscale objects which are referred to as nanostructures (Murmu and Adhikari 2010a). It is thus quite important to have proper knowledge of mechanical behavior for the development of nanostructures. Studying the behavior of structures at very small length scales has become one of new frontier of research in the area of computational nanomechanics (Ansari and Sahmani 2011). Structural elements such as beams, sheets, and plates in micro or nanolength scale, which are commonly used as components in microelectromechanical systems (MEMS) or nanoelectromechanical systems (NEMS) devices (Lu *et al.* 2007), present significant challenges to researchers in nanomechanics (Mahmoud *et al.* 2012). Invention of CNTs by Ijima in 1991 has initiated a new era in nanoworld (Danesh *et al.* 2012). Some of the excellent properties of CNTs are high stiffness, low density, very high aspect ratio, remarkable electronic properties, high conductivity, and high strength (Ehteshami and Hajabasi 2011). Some of the applications of CNTs include atomic force microscopes, field emitters, nanofillers for composite materials and nanoscale electronic devices. They are also used for the development of superconductive devices for MEMs and NEMs applications (Wang and Varadan 2006). Therefore, research on CNTs may contribute to some new applications (Wang and Varadan 2006). Hence, static and dynamic behaviors of CNTs have become one of the interesting topics in the past few years.

Similar to CNTs, nanoplates in the form of graphene sheets have aroused interest due to their unique superior properties. Graphene sheets are commonly used as components in MEMS/NEMS devices such as resonators, mass sensors and atomistic dust detectors (Farajpour *et al.* 2011). Nanoplates may be used as thin-film elements, nanosheet resonators, and paddle-like resonators (Aksencer and Aydogdu 2012).

Conducting experiments at nanoscale size is quite difficult and expensive. In this regard, development of appropriate mathematical models for nanostructures (such as graphene, CNT, nanorod, nanofiber, etc.) became an important concern (Pradhan and Phadikar 2009b). Generally, three approaches such as atomistic, hybrid atomistic–continuum mechanics, and continuum mechanics have been developed to model nanostructures (Narendar and Gopalakrishnan 2012). Some of the atomistic approaches are classical molecular dynamics, tight binding molecular dynamics, and density functional theory (Wang and Varadan 2006). But the atomic methods are limited to systems with a small number of molecules and atoms. As such, it is restricted to the study of small scale modeling (Wang and Varadan 2006). Also, this approach is computationally intensive and very expensive (Pradhan and Phadikar 2009a). Continuum mechanics results are found to be in good agreement with those obtained from atomistic and hybrid approaches (Narendar and Gopalakrishnan 2012).

Small scale nanotechnology makes the applicability of classical or local continuum models such as beam, shell, and plate questionable. Classical continuum models do not admit intrinsic size dependence in the elastic solutions of inclusions and inhomogeneities. At nanoscale size, the material microstructure such as lattice spacing between individual atoms becomes increasingly important and the discrete structure of the material can no longer be homogenized into a continuum. It is therefore needed to extend continuum models for considering scale effect in nanomaterial studies. As the length scales are reduced, the influences of long-range interatomic and intermolecular cohesive forces on the static and dynamic properties become significant and thus could not be neglected. Classical continuum mechanics exclude these effects and thus fail to capture the small scale effects when dealing with nanostructures. It is found that small size analysis using local theory overpredicts the results. Therefore, it is quite necessary to consider small effects for correct prediction of nanostructures (Narendar and Gopalakrishnan 2012). Small scale effects and the atomic forces must be incorporated in the realistic design of the nanostructures (viz., nanoresonators, nanoactuators, nanomachines, and nanooptomechanical systems) to achieve solutions with acceptable accuracy. Both experimental and atomistic simulation results show that when the dimensions of the structures become small, then the size effect has significant role in the mechanical properties (Murmu and Adhikari 2010a). Classical continuum models are scale-free theory and it does not include the effects arising from the small scale (Murmu and Adhikari 2010a). As such, various size-dependent continuum models such as strain gradient theory (Nix and Gao 1998), couple stress theory (Hadjesfandiari and Dargush 2011), modified couple stress theory (Asghari *et al.* 2010), and nonlocal elasticity theory (Eringen 1997) came into existence. Among these theories, nonlocal elasticity theory pioneered by Eringen has been widely used by the researchers (Thai 2012). Recent literature shows that nonlocal elasticity theory is being increasingly used for reliable and better analysis of nanostructures (Murmu and Adhikari 2010a). Applications of nonlocal continuum mechanics include lattice dispersion of elastic waves, fracture mechanics, dislocation mechanics, and wave propagation in composites (Aydogdu 2009). Generally, nonlocal elasticity theory is used in two forms: nonlocal differential elasticity and nonlocal integral elasticity (Ghannadpour *et al.* 2013). But nonlocal differential elasticity is more popular due to its simplicity (Ghannadpour *et al.* 2013). In nonlocal elasticity theory, the stress at a point is a function of the strains at all points in the domain, whereas in classical continuum models, the stress at a point is a function of the strains at that point in the domain (Murmu and Adhikari 2010a).

In the last few years, there has been extensive research on the bending, free vibration and buckling of nanostructures based on nonlocal elasticity theory. Vibration analysis of nanostructures include applications in structural engineering such as long span bridges, aerospace vehicles, automobiles, and many other industrial usages. When nanostructure elements are subjected to compressive in-plane loads, then these structures may buckle (Emam 2013). It is a well-known fact that buckling of a nanostructure initiates instability. In this context, knowledge of buckling load is quite necessary. As such, proper understanding of the stability response under in-plane loads for nanostructures is quite necessary.

Structural members with variable material properties are frequently used in engineering to satisfy various requirements. The literature reveals that previous studies done in nanobeams and nanoplates are mostly with constant parameters. But in actual practice, there may be a variation in these parameters. Hence, for practical applications of nanobeams and nanoplates, one should investigate geometrically nonlinearity model of nanostructures. Study of various aspects of nanotubes such as bending, buckling, thermal properties, etc. has attracted considerable attention among the researcher of nanotechnology. Thermal effects can induce an axial force within CNTs which may lead to bending and buckling (Lee and Chang 2009). Thermal vibration frequencies may be used to estimate Young's modulus of various nanotubes. Therefore, investigation of thermal effect has great importance. Moreover, the surrounding elastic medium such as Winkler-type and Pasternak-type elastic foundation has also a great influence on the analysis of CNTs.

The governing equation of beams and plates can be derived by using either vector mechanics or energy and variational principles. In this book, we have shown governing equations of various beam and plate theories based on nonlocal elasticity theory. At first, we have given overview of nonlocal elasticity theory below.

1.1.1 *Review of nonlocal elasticity theory*

According to nonlocal elasticity theory, the nonlocal stress tensor σ at a point x is expressed as (Murmu and Adhikari 2010a)

$$\sigma(x) = \int_V K(|x' - x|, \alpha)\tau \, dV(x'), \tag{1.1}$$

where τ is the classical stress tensor, $K(|x' - x|, \alpha)$ the nonlocal modulus, and $|x' - x|$ the Euclidean distance. The volume integral is taken over the region V occupied by the body. Here α is a material constant that depends on both internal and external characteristic lengths.

According to Hooke's law

$$\tau(x) = C(x) : \epsilon(x), \tag{1.2}$$

where C is the fourth-order elasticity tensor, ϵ the classical strain tensor, and : denotes double dot product.

Equation (1.1) is the integral constitutive relation which is quite difficult to solve. Hence equivalent differential form of this equation may be written as (Murmu and Adhikari 2010a)

$$(1 - \alpha^2 L^2 \nabla^2)\sigma = \tau, \quad \alpha = \frac{e_0 l_{int}}{L}, \tag{1.3}$$

where ∇^2 the Laplace operator, e_0 is a material constant which could be determined from experiments or by matching dispersion curves of plane waves with those of atomic lattice dynamics, l_{int} is an internal characteristic length such as lattice parameter, C–C bond length or granular distance, while L is an external characteristic length which is usually taken as the length of the nanostructure. The term $e_0 l_{int}$ is called the nonlocal parameter which reveals scale effect in models or it reveals the nanoscale effect on the response of structures.

1.2 Overview of Beam Theories

There are a number of beam theories in the literature. Below, we have discussed some of the beam theories.

The simplest beam theory is the Euler–Bernoulli beam theory (EBT). Beam theories such as EBT, Timoshenko beam theory (TBT), Reddy–Bickford beam theory (RBT), general exponential shear deformation beam theory (ABT), Levinson beam theory (LBT) and plate theories such as classical plate theory (CPT), first-order shear deformation plate theory (FSDT), third-order shear deformation plate theory (TSDT), etc. have been developed by various researchers (Wang *et al.* 2000).

In EBT, both transverse shear and transverse normal strains are neglected. In TBT, a constant state of transverse shear strain and also shear stress with respect to the thickness coordinate is included. Due to this constant shear stress assumption, shear correction factors are needed to compensate the error. Shear correction factors depend on the material and geometric parameters, the loading and boundary conditions. Next, we have different third-order beam theories such as RBT, LBT, etc. In these theories, shear correction factors are not needed (Wang *et al.* 2000). Levinson derived equations of equilibrium using vector approach. As such, governing equations are same as those of TBT. But Reddy derived equations of motion using the principle of virtual displacements. Thus, Reddy (1997) and LBTs have same displacement and strain fields with different equations of motion. One may also note that the equations of Levinson's beam theory cannot be derived from the principle of total potential energy (Reddy 2007).

1.2.1 *Beam theories*

In this section, we have discussed displacement fields, energies of the system and governing equations of four types of beam theories, viz. EBT, TBT, RBT, and LBT. A schematic diagram for nanobeams embedded within an elastic medium characterized by spring constant K_w and shear constant K_g has been shown in Fig. 1.1.

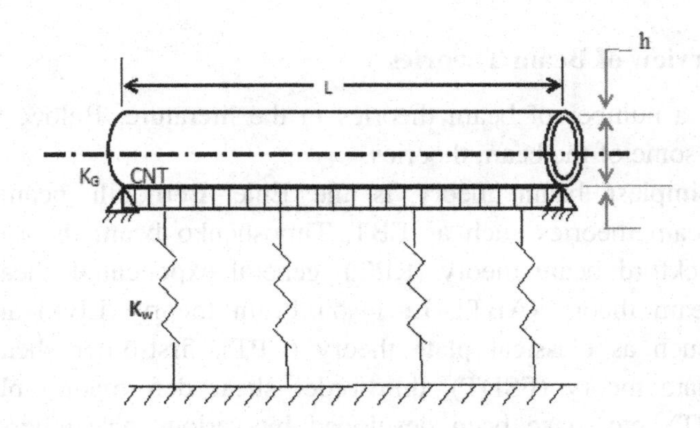

Fig. 1.1 Single-walled CNT embedded within an elastic medium

In all the beam theories (Reddy 2007), x-, y-, and z-coordinates are taken along the length, width, and thickness (the height) of the beam, respectively. All applied loads and geometry are such that the displacements (u_1, u_2, u_3) along the coordinates (x, y, z) are only functions of x- and z-coordinates as well as time t. It may be noted here that we have not considered axial displacement of the point $(x, 0)$ on the mid-plane $(z = 0)$ of the beam. One may note that the notation M, P, Q, and R which will be used in our subsequent paragraphs are defined as below:

$$R = \int_A z^2 \sigma_{xz}\, dA.$$

1.2.1.1 *Euler–Bernoulli beam theory (EBT)*

Based on EBT, the displacement fields are given by (Reddy 2007)

$$\begin{aligned} u_1 &= -z\frac{\partial w}{\partial x}, \\ u_2 &= 0, \\ u_3 &= w(x, t), \end{aligned} \tag{1.4}$$

where (u_1, u_2, u_3) are the displacements along x-, y-, and z-coordinates, respectively, and w is the transverse displacement of the point $(x, 0)$ on the mid-plane $(z = 0)$ of the beam.

The only non-zero strain of the EBT is written as

$$\varepsilon_{xx} = -z\frac{\partial^2 w}{\partial x^2}. \tag{1.5}$$

Governing equation of Euler–Bernoulli nanobeams may be written as (Reddy 2007)

$$\frac{\partial^2 M}{\partial x^2} + q - \bar{N}\frac{\partial^2 w}{\partial x^2} = m_0\frac{\partial^2 w}{\partial t^2}, \tag{1.6}$$

where q is the transverse force per unit length, $M = \int_A z\sigma_{xx}\, dA$, $\bar{N}$ the applied axial compressive force and m_0 is the mass inertia defined by $m_0 = \int_A \rho\, dA = \rho A$, with A being the cross-sectional area of the beam.

According to Eringen's nonlocal theory (Eringen 1987), the constitutive relation for Euler–Bernoulli nanobeam is given by (Reddy 2007)

$$M - \mu\frac{\partial^2 M}{\partial x^2} = -EI\frac{\partial^2 w}{\partial x^2}, \tag{1.7}$$

where E is Young's modulus and I is the second moment of area about y-axis. It may be noted here that $\mu = (e_0 l_{\text{int}})^2$ is the nonlocal parameter, where e_0 and l_{int} denote material constant and internal characteristic length, respectively.

Using Eqs. (1.6) and (1.7), the nonlocal form of M may be written as

$$M = -EI\frac{\partial^2 w}{\partial x^2} + \mu\left(\bar{N}\frac{\partial^2 w}{\partial x^2} - q + m_0\frac{\partial^2 w}{\partial t^2}\right). \tag{1.8}$$

Governing equation in terms of displacement is rewritten as

$$\begin{aligned} -EI\frac{\partial^4 w}{\partial x^4} + \mu\frac{\partial^2}{\partial x^2}\left[\bar{N}\frac{\partial^2 w}{\partial x^2} - q + m_0\frac{\partial^2 w}{\partial t^2}\right] \\ +q - \bar{N}\frac{\partial^2 w}{\partial x^2} = m_0\frac{\partial^2 w}{\partial t^2}. \end{aligned} \tag{1.9}$$

1.2.1.2 *Timoshenko beam theory (TBT)*

The displacement fields are based on (Reddy 2007)

$$\begin{aligned} u_1 &= z\phi(x,t), \\ u_2 &= 0, \\ u_3 &= w(x,t), \end{aligned} \tag{1.10}$$

where ϕ is the rotation of the cross-section.

The non-zero strains of the TBT are given by

$$\epsilon_{xx} = z\frac{\partial \phi}{\partial x}, \tag{1.11}$$

$$\gamma_{xz} = \phi + \frac{\partial w}{\partial x}. \tag{1.12}$$

Constitutive relations for TBT may be written as (Reddy 2007)

$$M - \mu\frac{\partial^2 M}{\partial x^2} = EI\frac{\partial \phi}{\partial x}, \tag{1.13}$$

$$Q - \mu\frac{\partial^2 Q}{\partial x^2} = k_s GA\left(\phi + \frac{\partial w}{\partial x}\right), \tag{1.14}$$

where G is the shear modulus, $Q = \int_A \sigma_{xz}\, dA$ and k_s is the shear correction factor.

Governing equations of this beam theory are (Reddy 2007)

$$\frac{\partial M}{\partial x} - Q = m_2 \frac{\partial^2 \phi}{\partial t^2}, \tag{1.15}$$

$$\frac{\partial Q}{\partial x} + q - \bar{N}\frac{\partial^2 w}{\partial x^2} = m_0 \frac{\partial^2 w}{\partial t^2}. \tag{1.16}$$

Using Eqs. (1.13)–(1.16), the nonlocal form of M and Q may be obtained as

$$M = EI\frac{\partial \phi}{\partial x} + \mu\left[-q + \bar{N}\frac{\partial^2 w}{\partial x^2} + m_0\frac{\partial^2 w}{\partial t^2} + m_2\frac{\partial^3 \phi}{\partial x \partial t^2}\right], \tag{1.17}$$

$$Q = GAk_s\left(\phi + \frac{\partial w}{\partial x}\right) + \mu\frac{\partial}{\partial x}\left[-q + \bar{N}\frac{\partial^2 w}{\partial x^2} + m_0\frac{\partial^2 w}{\partial t^2}\right]. \tag{1.18}$$

Using Eqs. (1.17) and (1.18) in Eqs. (1.15) and (1.16), we have the governing equations as

$$\begin{aligned} GAk_s\left(\frac{\partial \phi}{\partial x} + \frac{\partial^2 w}{\partial x^2}\right) + q - \bar{N}\frac{\partial^2 w}{\partial x^2} - \mu\left[\frac{\partial^2 q}{\partial x^2} - \bar{N}\frac{\partial^4 w}{\partial x^4}\right] \\ = m_0\left(\frac{\partial^2 w}{\partial t^2} - \mu\frac{\partial^4 w}{\partial x^2 \partial t^2}\right), \end{aligned} \tag{1.19}$$

$$EI\frac{\partial^2 \phi}{\partial x^2} - GAk_s\left(\phi + \frac{\partial w}{\partial x}\right) = m_2\frac{\partial^2 \phi}{\partial t^2} - \mu m_2\frac{\partial^4 \phi}{\partial x^2 \partial t^2}. \tag{1.20}$$

1.2.1.3 *Reddy beam theory (RBT)*

Displacement fields for RBT are based on (Reddy 2007)

$$\begin{aligned} u_1 &= z\phi(x,t) - c_1 z^3\left(\phi + \frac{\partial w}{\partial x}\right), \\ u_2 &= 0, \\ u_3 &= w(x,t), \end{aligned} \tag{1.21}$$

where $c_1 = 4/3h^2$, with h being the height of the beam.

The non-zero strains of the RBT are

$$\varepsilon_{xx} = z(1 - c_1 z^2)\frac{\partial \phi}{\partial x} - c_1 z^3 \frac{\partial^2 w}{\partial x^2},$$
$$\gamma_{xz} = (1 - c_2 z^2)\left(\frac{\partial w}{\partial x} + \phi\right), \tag{1.22}$$

where $c_2 = 4/h^2$.

Nonlocal constitutive equations take the following form in the case of RBT:

$$\hat{M} - \mu \frac{\partial^2 \hat{M}}{\partial x^2} = E\hat{I}\frac{\partial \phi}{\partial x} + E\hat{J}(-c_1)\left(\frac{\partial \phi}{\partial x} + \frac{\partial^2 w}{\partial x^2}\right),$$
$$\hat{Q} - \mu \frac{\partial^2 \hat{Q}}{\partial x^2} = G\bar{A}\left(\phi + \frac{\partial w}{\partial x}\right) + G\bar{I}(-c_2)\left(\phi + \frac{\partial w}{\partial x}\right), \tag{1.23}$$
$$P - \mu \frac{\partial^2 P}{\partial x^2} = EJ\frac{\partial \phi}{\partial x} + EK(-c_1)\left(\frac{\partial \phi}{\partial x} + \frac{\partial^2 w}{\partial x^2}\right),$$

where I, J, and K are the second-, fourth-, and sixth-order moments of area, respectively, about the y-axis and are defined as $(I, J, K) = \int_A (z^2, z^4, z^6)\, dA$. Here $P = \int_A z^3 \sigma_{xx}\, dA$.

Also $\hat{M} = M - c_1 P$, $\hat{Q} = Q - c_2 R$, $\hat{I} = I - c_1 J$, $\hat{J} = J - c_1 K$, $\bar{A} = A - c_2 I$, $\bar{I} = I - c_2 J$, $\tilde{A} = \bar{A} - c_2 \bar{I}$.

$\hat{M}$ and $\hat{Q}$ are given by (Reddy 2007)

$$\hat{M} = E\hat{I}\frac{\partial \phi}{\partial x} - c_1 E\hat{J}\left(\frac{\partial \phi}{\partial x} + \frac{\partial^2 w}{\partial x^2}\right)$$
$$+ \mu\left[-c_1 \frac{\partial^2 P}{\partial x^2} - q + \bar{N}\frac{\partial^2 w}{\partial x^2} + m_0 \frac{\partial^2 w}{\partial t^2}\right], \tag{1.24}$$

$$\hat{Q} = G\tilde{A}\left(\phi + \frac{\partial w}{\partial x}\right) + \mu\left[-c_1 \frac{\partial^3 P}{\partial x^3} + \bar{N}\frac{\partial^3 w}{\partial x^3} - \frac{\partial q}{\partial x}\right]$$
$$+ \mu m_0 \frac{\partial^3 w}{\partial x \partial t^2}. \tag{1.25}$$

As such, governing equations may be written as (Reddy 2007)

$$\begin{aligned}
&G\tilde{A}\left(\frac{\partial \phi}{\partial x}+\frac{\partial^2 w}{\partial x^2}\right)-\bar{N}\frac{\partial^2 w}{\partial x^2}+q+\mu\left[\bar{N}\frac{\partial^4 w}{\partial x^4}-\frac{\partial^2 q}{\partial x^2}\right]\\
&\quad+c_1 EJ\frac{\partial^3 \phi}{\partial x^3}-c_1^2 EK\left(\frac{\partial^3 \phi}{\partial x^3}+\frac{\partial^4 w}{\partial x^4}\right)\\
&\quad=m_0\left(\frac{\partial^2 w}{\partial t^2}-\mu\frac{\partial^4 w}{\partial x^2 \partial t^2}\right), \qquad (1.26)
\end{aligned}$$

$$E\hat{I}\frac{\partial^2 \phi}{\partial x^2}-c_1 E\hat{J}\left(\frac{\partial^2 \phi}{\partial x^2}+\frac{\partial^3 w}{\partial x^3}\right)-G\tilde{A}\left(\phi+\frac{\partial w}{\partial x}\right)=0. \qquad (1.27)$$

It may be noted here that in our problems, we have neglected the higher-order inertias, i.e. m_2, m_4, and m_6.

1.2.1.4 *Levinson beam theory (LBT)*

LBT is based on the following displacement fields (Reddy 2007):

$$\begin{aligned}
u_1&=z\phi(x,t)-c_1 z^3\left(\phi+\frac{\partial w}{\partial x}\right),\\
u_2&=0, \qquad (1.28)\\
u_3&=w(x,t).
\end{aligned}$$

It may be noted that displacement and strain fields of the LBT is the same as that of RBT.

Governing equations are given as follows (Reddy 2007):

$$\begin{aligned}
\frac{\partial Q}{\partial x}+q-\bar{N}\frac{\partial^2 w}{\partial x^2}&=m_0\frac{\partial^2 w}{\partial t^2},\\
\frac{\partial M}{\partial x}-Q&=m_2\frac{\partial^2 \phi}{\partial t^2}.
\end{aligned} \qquad (1.29)$$

Nonlocal constitutive relations may be expressed as (Reddy 2007)

$$\begin{aligned}
M-\mu\frac{\partial^2 M}{\partial x^2}&=EI\frac{\partial \phi}{\partial x}+EJ(-c_1)\left(\frac{\partial \phi}{\partial x}+\frac{\partial^2 w}{\partial x^2}\right),\\
Q-\mu\frac{\partial^2 Q}{\partial x^2}&=GA\left(\phi+\frac{\partial w}{\partial x}\right)+GI(-c_2)\left(\phi+\frac{\partial w}{\partial x}\right).
\end{aligned} \qquad (1.30)$$

Using Eqs. (1.29) and (1.30), M and Q are obtained as

$$M = EI\frac{\partial \phi}{\partial x} - c_1 EJ\left(\frac{\partial \phi}{\partial x} + \frac{\partial^2 w}{\partial x^2}\right) + \mu\left[-q + \bar{N}\frac{\partial^2 w}{\partial x^2} + m_0\frac{\partial^2 w}{\partial t^2} + m_2\frac{\partial^3 \phi}{\partial x \partial t^2}\right], \tag{1.31}$$

$$Q = G\bar{A}\left(\phi + \frac{\partial w}{\partial x}\right) + \mu\left[\bar{N}\frac{\partial^3 w}{\partial x^3} - \frac{\partial q}{\partial x} + m_0\frac{\partial^3 w}{\partial x \partial t^2}\right]. \tag{1.32}$$

Utilizing Eqs. (1.31) and (1.32), governing differential equations may be transformed to

$$G\bar{A}\left(\frac{\partial \phi}{\partial x} + \frac{\partial^2 w}{\partial x^2}\right) + q - \bar{N}\frac{\partial^2 w}{\partial x^2} + \mu\left[\bar{N}\frac{\partial^4 w}{\partial x^4} - \frac{\partial^2 q}{\partial x^2}\right] = m_0\left(\frac{\partial^2 w}{\partial t^2} - \mu\frac{\partial^4 w}{\partial^2 x \partial t^2}\right), \tag{1.33}$$

$$EI\frac{\partial^2 \phi}{\partial x^2} - c_1 EJ\left(\frac{\partial^2 \phi}{\partial x^2} + \frac{\partial^3 w}{\partial x^3}\right) - G\bar{A}\left(\phi + \frac{\partial w}{\partial x}\right) = m_2\left(\frac{\partial^2 \phi}{\partial t^2} - \mu\frac{\partial^4 \phi}{\partial^2 x \partial t^2}\right). \tag{1.34}$$

1.3 Overview of Plate Theories

There are a number of plate theories in the literature, and a review of these theories is given by Reddy (1997). The simplest plate theory is the CPT. In CPT, both transverse shear and transverse normal strains are neglected. The next theory in the hierarchy of refined theories is the FSDT. In FSDT, transverse shear strain is assumed to be constant with respect to the thickness coordinate. As such, shear correction factors are taken into consideration to compensate the error. The shear correction factors depend not only on the geometric parameters, but also on the loading and boundary conditions of the plate. We may also find a number of third-order plate theories in the literature (Reddy 1997). Similar to third-order beam theories, shear correction factors are not needed in third-order plate theories (Wang *et al.* 2000).

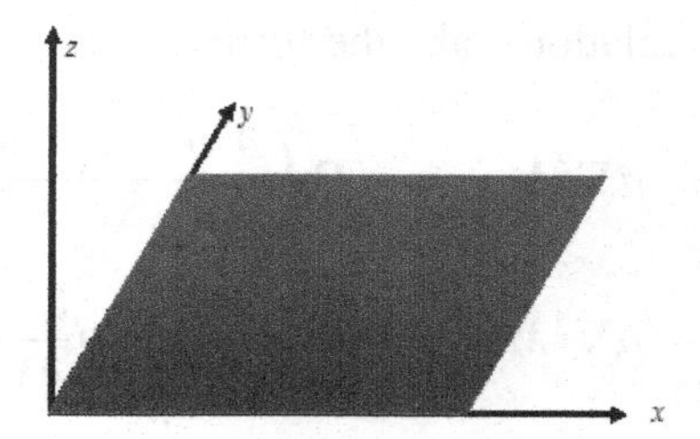

Fig. 1.2 Schematic of a single-layered nanoplate

1.3.1 *Plate theory*

In this section, we have shown some of the preliminaries related to CPT and FSDT. Figure 1.2 shows the coordinate system used for the nanoplate. The x-, y-, and z-coordinates are taken along the length, width, and thickness of the plate, respectively, and origin is chosen at one corner of the mid-plane of the plate.

1.3.1.1 *Classical plate theory (CPT)*

Based on CPT, the displacement fields (u_1, u_2, u_3) at time t are written as (Pradhan and Phadikar 2009b)

$$
\begin{aligned}
u_1 &= u(x, y, t) - z\frac{\partial w}{\partial x}, \\
u_2 &= v(x, y, t) - z\frac{\partial w}{\partial y}, \\
u_3 &= w(x, y, t),
\end{aligned} \tag{1.35}
$$

where u, v, and w denote displacement of the point $(x, y, 0)$ along x, y, and z directions, respectively.

The strain components are expressed as

$$
\varepsilon_{xx} = \frac{\partial u}{\partial x} - z\frac{\partial^2 w}{\partial x^2}, \quad \varepsilon_{yy} = \frac{\partial v}{\partial x} - z\frac{\partial^2 w}{\partial y^2},
$$

$$
\varepsilon_{xy} = \frac{1}{2}\left(\frac{\partial u}{\partial y} + \frac{\partial v}{\partial x} - 2z\frac{\partial^2 w}{\partial x \partial y}\right), \quad \varepsilon_{zz} = \varepsilon_{xz} = \varepsilon_{yz} = 0.
$$

Nonlocal constitutive relations take the form

$$\begin{aligned} M_{xx} - \mu\nabla^2 M_{xx} &= -D\left(\frac{\partial^2 w}{\partial x^2} + \nu\frac{\partial^2 w}{\partial y^2}\right), \\ M_{yy} - \mu\nabla^2 M_{yy} &= -D\left(\frac{\partial^2 w}{\partial y^2} + \nu\frac{\partial^2 w}{\partial x^2}\right), \\ M_{xy} - \mu\nabla^2 M_{xy} &= -D(1-\nu)\frac{\partial^2 w}{\partial x \partial y}, \end{aligned} \tag{1.36}$$

where (M_{xx}, M_{yy}, M_{xy}) are moment resultants, ∇^2 the Laplacian operator in 2D Cartesian coordinate system and $D = Eh^3/12(1-\nu^2)$ denotes bending rigidity of the plate.

In the expression of bending rigidity, h is the thickness of the plate, E Young's modulus, and ν is Poisson's ratio.

Governing equation in terms of the displacement is written as (Pradhan and Phadikar 2009b)

$$\begin{aligned} -D\nabla^4 w + (1-\mu\nabla^2)\Bigg[q + N_{xx}\frac{\partial^2 w}{\partial x^2} + N_{yy}\frac{\partial^2 w}{\partial y^2} + 2N_{xy}\frac{\partial^2 w}{\partial x \partial y} \\ - m_0\frac{\partial^2 w}{\partial t^2} + m_2\left(\frac{\partial^4 w}{\partial x^2 \partial t^2} + \frac{\partial^4 w}{\partial y^2 \partial t^2}\right)\Bigg] = 0, \end{aligned} \tag{1.37}$$

where q is the transverse distributed load and (N_{xx}, N_{xy}, N_{yy}) are in-plane force resultants. Also m_0 and m_2 are mass moments of inertia which are defined as

$m_0 = \int_{-h/2}^{h/2} \rho \, dz$, $m_2 = \int_{-h/2}^{h/2} \rho h^2 \, dz$, with ρ being the density of the material.

1.3.1.2 *First-order shear deformation plate theory (FSDT)*

Based on this plate theory, the displacement fields (u_1, u_2, u_3) at time t are written as (Pradhan and Phadikar 2009b)

$$\begin{aligned} u_1 &= u(x, y, t) + z\psi_x(x, y, t), \\ u_2 &= v(x, y, t) + z\psi_y(x, y, t), \\ u_3 &= w(x, y, t), \end{aligned} \tag{1.38}$$

where u, v, and w denote displacement of the point $(x, y, 0)$ along x, y, and z directions, respectively. Here ψ_x and ψ_y are the rotations of a transverse normal in the plate with respect to x- and y-axis, respectively.

The strain components are expressed as (Pradhan and Phadikar 2009b)

$$\varepsilon_{xx} = \frac{\partial u}{\partial x} + z\frac{\partial \psi_x}{\partial x}, \quad \varepsilon_{yy} = \frac{\partial v}{\partial y} + z\frac{\partial \psi_y}{\partial y}, \quad \varepsilon_{yz} = \frac{1}{2}\left(\frac{\partial w}{\partial y} + \psi_y\right),$$

$$\varepsilon_{xz} = \frac{1}{2}\left(\frac{\partial w}{\partial x} + \psi_x\right), \quad \varepsilon_{xy} = \frac{1}{2}\left(\frac{\partial u}{\partial y} + z\frac{\partial v}{\partial x} + z\frac{\partial \psi_x}{\partial y} + \frac{\partial \psi_y}{\partial x}\right),$$

$$\varepsilon_{zz} = 0.$$

Nonlocal constitutive relations take the form

$$M_1^{xx} - \mu\nabla^2 M_1^{xx} = D\left(\frac{\partial \psi_x}{\partial x} + \nu\frac{\partial \psi_y}{\partial y}\right),$$

$$M_1^{yy} - \mu\nabla^2 M_1^{yy} = D\left(\frac{\partial \psi_y}{\partial y} + \nu\frac{\partial \psi_x}{\partial x}\right),$$

$$M_1^{xy} - \mu\nabla^2 M_1^{xy} = \frac{1}{2}D(1-\nu)\left(\frac{\partial \psi_x}{\partial y} + \nu\frac{\partial \psi_y}{\partial x}\right), \tag{1.39}$$

$$V_0^{xx} - \mu\nabla^2 V_0^{xx} = K_s G h\left(\psi_x + \frac{\partial w}{\partial x}\right),$$

$$V_0^{yy} - \mu\nabla^2 V_0^{yy} = K_s G h\left(\psi_y + \frac{\partial w}{\partial y}\right),$$

where $M_1^{xx} = \int_{-h/2}^{h/2} z\sigma_{xx}^{nl}\,dz$, $M_1^{yy} = \int_{-h/2}^{h/2} z\sigma_{yy}^{nl}\,dz$, $M_1^{xy} = \int_{-h/2}^{h/2} z\sigma_{xy}^{nl}\,dz$, $V_0^{xx} = \int_{-h/2}^{h/2} \sigma_{xz}^{nl}\,dz$, $V_0^{yy} = \int_{-h/2}^{h/2} z\sigma_{yz}^{nl}\,dz$. Here h denotes the height of the plate, G the shear modulus of the plate material, K_s denotes the shear correction factor, σ_{xx}^{nl}, σ_{yy}^{nl}, σ_{zz}^{nl}, σ_{xy}^{nl}, σ_{yz}^{nl}, and σ_{xz}^{nl} represent the nonlocal stress tensors.

Governing equation in terms of the displacement is written as (Pradhan and Phadikar 2009b)

$$m_0\left(\frac{\partial^2 w}{\partial t^2} - \mu\nabla^2\frac{\partial^2 w}{\partial t^2}\right)$$

$$= K_s G h\left(\frac{\partial \psi_x}{\partial x} + \frac{\partial \psi_y}{\partial y} + \frac{\partial^2 w}{\partial x^2} + \frac{\partial^2 w}{\partial y^2}\right)$$

$$+ (1 - \mu \nabla^2)\left[q + N_0^{xx}\frac{\partial^2 w}{\partial x^2} + N_0^{yy}\frac{\partial^2 w}{\partial y^2} + 2N_0^{xy}\frac{\partial^2 w}{\partial x \partial y}\right],$$

$$D\frac{\partial^2 \psi_x}{\partial x^2} + \frac{1}{2}(1-\nu)\frac{\partial^2 \psi_x}{\partial y^2} + \frac{1}{2}(1+\nu)\frac{\partial^2 \psi_y}{\partial x \partial y} - K_s G h\left(\psi_x + \frac{\partial w}{\partial x}\right)$$

$$= m_2\left(\frac{\partial^2 \psi_x}{\partial t^2} - \mu \nabla^2 \frac{\partial^2 \psi_x}{\partial t^2}\right),$$

$$D\frac{\partial^2 \psi_y}{\partial y^2} + \frac{1}{2}(1-\nu)\frac{\partial^2 \psi_y}{\partial x^2} + \frac{1}{2}(1+\nu)\frac{\partial^2 \psi_y}{\partial x \partial y} - K_s G h\left(\psi_y + \frac{\partial w}{\partial y}\right)$$

$$= m_2\left(\frac{\partial^2 \psi_y}{\partial t^2} - \mu \nabla^2 \frac{\partial^2 \psi_y}{\partial t^2}\right). \tag{1.40}$$

Here q is the transverse distributed load, $N_0^{xx} = \int_{-h/2}^{h/2} \sigma_{xx}^{nl}\, dz$, $N_0^{yy} = \int_{-h/2}^{h/2} \sigma_{yy}^{nl}\, dz$, and $N_0^{xy} = \int_{-h/2}^{h/2} \sigma_{xy}^{nl}\, dz$.

Chapter 2

Analytical Methods

In this chapter, we consider exact solutions of bending, buckling, and vibration of simply supported beam and plates.

The boundary conditions of simply supported beams are $w = 0$ and $M = 0$ at $x = 0, L$.

We assume the generalized displacements as

$$w(x,t) = \sum_{n=1}^{\infty} W_n \sin \frac{n\Pi x}{L} e^{i\omega_n t}, \tag{2.1}$$

$$\phi(x,t) = \sum_{n=1}^{\infty} \varphi_n \cos \frac{n\Pi x}{L} e^{i\omega_n t}. \tag{2.2}$$

In the case of bending, we set $\bar{N}$ and all time derivatives to zero and simultaneously we assume that the distributed load is of the form

$$\begin{aligned} q(x) &= \sum_{n=1}^{\infty} Q_n \sin \frac{n\Pi x}{L}, \\ Q_n &= \frac{2}{L} \int_0^L q(x) \sin \frac{n\Pi x}{L} dx. \end{aligned} \tag{2.3}$$

Here the coefficients Q_n associated with various types of loads are given below:

$$q(x) = q_0, \quad Q_n = \frac{4q_0}{n\Pi}, \quad n = 1, 3, 5, \ldots$$

$$q(x) = \frac{q_0 x}{L}, \quad Q_n = \frac{2q_0}{n\Pi}(-1)^{n+1}, \quad n = 1, 2, 3, \ldots.$$

Similarly, one may find the coefficients for other types of loads in Reddy (2007).

For buckling, we set q and all time derivatives to zero and for free vibration, we set q and $\bar{N}$ to zero.

2.1 Euler–Bernoulli Beam Theory (EBT)

Substituting Eqs. (2.1) and (2.3) in Eq. (1.9), we obtain

$$\left\{\lambda_n\left[\bar{N}\left(\frac{n\Pi}{L}\right)^2+\omega_n^2\left(m_0+m_2\left(\frac{n\Pi}{L}\right)^2\right)\right]-EI\left(\frac{n\Pi}{L}\right)^4\right\} W_n+\lambda_n Q_n=0, \quad (2.4)$$

where $\lambda_n=1+\mu\left(\frac{n\Pi}{L}\right)^2$ for any n.

Bending: The static deflection is given by setting $\bar{N}$ and ω_n^2 to zero. As such, we have

$$w(x)=\sum_{n=1}^{\infty}\frac{\lambda_n Q_n L^4}{n^4\Pi^4 EI}\sin\frac{n\Pi x}{L}. \quad (2.5)$$

It may be conclude that the nonlocal parameter λ_n has the effect of increasing the deflection.

Buckling: For buckling solutions, we have

$$\left\{\lambda_n\left[\bar{N}\left(\frac{n\Pi}{L}\right)^2+\omega_n^2\left(m_0+m_2\left(\frac{n\Pi}{L}\right)^2\right)\right]-\right\}=EI\left(\frac{n\Pi}{L}\right)^4. \quad (2.6)$$

In particular, the critical buckling load is obtained by setting $\omega_n=0$ and $n=1$ in the above equation.

$$\bar{N}=\frac{1}{\lambda_1}\frac{\Pi^2 EI}{L^2}.$$

It is observed that the nonlocal parameter λ_n has the effect of reducing the critical buckling load.

Vibration: The natural frequencies are given by

$$\omega_n^2 = \frac{1}{\lambda_n M_n}\left(\frac{n\Pi}{L}\right)^4 EI, \quad M_n = m_0 + m_2\left(\frac{n\Pi}{L}\right)^2.$$

Both rotary inertia m_2 as well as the nonlocal parameter λ_n have the effect of decreasing frequencies of vibration.

2.2 Timoshenko Beam Theory (TBT)

Substituting Eqs. (2.1)–(2.3) in Eqs. (1.19) and (1.20), we obtain

$$\begin{aligned}
&-GAK_s\left(\frac{n\Pi}{L}\right)\left(\varphi_n + \frac{n\Pi}{L}W_n\right) + \lambda_n Q_n \\
&+ \lambda_n \bar{N}^T\left(\frac{n\Pi}{L}\right)^2 W_n + \lambda_n m_0 \omega_n^2 W_n = 0, \\
&-EI\left(\frac{n\Pi}{L}\right)^2 \varphi_n - GAK_s\left(\varphi_n + \frac{n\Pi}{L}W_n\right) \\
&+ \lambda_n m_2 \omega_n^2 \varphi_n = 0.
\end{aligned} \tag{2.7}$$

Bending: For static bending, we obtain

$$\begin{aligned}
w(x) &= \sum_{n=1}^{\infty} \lambda_n \Lambda_n \frac{Q_n L^4}{n^4\Pi^4 EI}\sin\frac{n\Pi x}{L}, \\
\phi(x) &= \sum_{n=1}^{\infty} \lambda_n \Lambda_n \frac{Q_n L^3}{n^3\Pi^3 EI}\cos\frac{n\Pi x}{L},
\end{aligned} \tag{2.8}$$

where $\Lambda_n = (1 + n^2\Pi^2\Omega)$, $\Omega = EI/K_s GAL^2$.

Buckling: The critical buckling load is given by

$$\bar{N} = \frac{1}{\lambda_1 \Lambda_1}\frac{\Pi^2 EI}{L^2}.$$

Vibration: The natural frequencies ω_n^2 can be computed from

$$\frac{m_0 m_2}{GAK_s}\lambda_n^2\omega_n^4 - \left[m_0\Lambda_n + m_2\left(\frac{n\Pi}{L}\right)^2\right]\lambda_n\omega_n^2 + EI\left(\frac{n\Pi}{L}\right)^4 = 0. \tag{2.9}$$

If we neglect the first term, we obtain

$$\omega_n^2 = \frac{1}{\lambda_n \Sigma_n}\left(\frac{n\Pi}{L}\right)^4 EI, \quad \Sigma_n = m_0\Lambda_n + m_2\left(\frac{n\Pi}{L}\right)^2.$$

2.3 Solution using Navier's Approach

It may be noted that Navier's approach may be applied for simply supported plate.

The simply supported boundary conditions for classical plate theory (CPT) may be written as

at $x = 0$ and $x = a$,

$$u = 0, \quad v = 0, \quad N_0^{xx} = 0, \quad M_1^{xx} = 0.$$

At $y = 0$ and $y = b$,

$$u = 0, \quad v = 0, \quad N_0^{yy} = 0, \quad M_1^{yy} = 0.$$

Next, let us assume the generalized displacements as

$$w(x,t) = \sum_{m=1}^{\infty}\sum_{n=1}^{\infty} W_{mn}\sin(\alpha x)\sin(\beta y)e^{i\omega_{mn}t}, \tag{2.10}$$

where $\alpha = m\Pi/a$ and $\beta = n\Pi/b$.

It is assumed that the plate is free from any in-plane or transverse loadings. So we have, $N_0^{xx} = N_0^{yy} = N_0^{xy} = q = 0$.

Substituting Eq. (2.10) into Eq. (1.37), we get

$$-D(\alpha^2+\beta^2)^2 W_{mn}^c = -M_{mn}\lambda_{mn}\omega_{mn}^c W_{mn}^c, \tag{2.11}$$

where $\lambda_{mn} = 1 + \mu(\alpha^2+\beta^2)$.

Using Eq. (2.11), we obtain natural frequencies as

$$\omega_{mn}^c = \sqrt{\frac{D(\alpha^2+\beta^2)^2}{M_{mn}\lambda_{mn}}},$$

where $M_{mn} = m_0 + m_2(\alpha^2+\beta^2)$.

It is seen that increase in nonlocal parameter would decrease the natural vibration frequencies.

Chapter 3

Numerical Methods

Two important numerical methods such as Rayleigh–Ritz and differential quadrature have been applied to investigate bending, buckling, and vibration of nanobeams and nanoplates. The novelty of the methods is that boundary conditions may easily be handled. In this chapter, we have discussed elaborately about these methods.

3.1 Rayleigh–Ritz Method

Rayleigh–Ritz method has been implemented in the bending, buckling, and free vibration of nanobeams based on Euler–Bernoulli beam theory (EBT) and Timoshenko beam theory (TBT). We have also applied Rayleigh–Ritz method in the bending, buckling, and free vibration of nanoplates based on classical plate theory (CPT). As such, we have discussed Rayleigh–Ritz method for the above-said problems. One may note that subsequent notation have been defined in the preliminaries of Chapter 1.

3.1.1 *Bending problems*

We have investigated bending of nanobeams based on EBT and TBT. Boundary characteristic orthogonal polynomials have used as shape functions in the Rayleigh–Ritz method. As such, we have discussed below the procedure of application of Rayleigh–Ritz method in the bending of nanobeams.

3.1.1.1 *Euler–Bernoulli beam theory (EBT)*

The strain energy U may be given as (Wang *et al.* 2000)

$$U = -\frac{1}{2}\int_0^L M\frac{d^2w}{dx^2}dx, \tag{3.1}$$

where M may be obtained from Eq. (1.8) by setting $\bar{N}$ and all time derivatives to zero. As such,

$$M = -EI\frac{d^2w}{dx^2} - \mu q. \tag{3.2}$$

The potential energy of the transverse force q may be given as (Ghannadpour *et al.* 2013)

$$V = -\frac{1}{2}\int_0^L qw\, dx. \tag{3.3}$$

The total energy U_T of the system is written as

$$U_T = \frac{1}{2}\int_0^L \left(EI\left(\frac{d^2w}{dx^2}\right)^2 + \mu q\frac{d^2w}{dx^2} - qw\right)dx. \tag{3.4}$$

It is important to note that in all of the problems in subsequent chapters, the domain has been transformed from $[0, L]$ to $[0, 1]$ by using the relation $X = x/L$. As such, we discussed the methods taking the domain as $[0, 1]$.

To apply Rayleigh–Ritz method, the displacement function (w) can be represented in the form of a series

$$w(X) = \sum_{k=1}^{n} c_k\varphi_k, \tag{3.5}$$

where n is the number of terms needed in the series and c_k's are unknowns. Here φ_k are polynomials which are consisting of a boundary polynomial specifying support conditions (essential boundary conditions) multiplied by one-dimensional simple polynomials. That is, $\varphi_k = ff_k$ with $f_k = X^{k-1}$, $k = 1, 2, \ldots, n$ and $f = X^u(1-X)^v$, where u and v take the values 0, 1, and 2, respectively, for free (F), simply supported (S), and clamped (C) edge conditions.

If instead of simple polynomials, Chebyshev polynomials (T_k) are used, then $\varphi_k = fT_k$. These Chebyshev polynomials are well known and we have

$T_0 = 1, T_1 = X$, and then T_{n+1} may be obtained by the recurrence relation $T_{(n+1)}(X) = 2XT_n(X) - T_{n-1}(X)$.

When φ_k are orthogonal polynomials, then Gram–Schmidt process as discussed below may be used to obtain these polynomials from a set of linearly independent functions $F_k = ff_k$. Then orthonormal polynomials $\hat{\varphi_k}$ may be obtained from orthogonal polynomials φ_k if we divide φ_k by the norm of φ_k. Below, we have included detailed procedure to obtain orthogonal and orthonormal polynomials.

$$\varphi_1 = F_1, \quad \varphi_k = F_k - \sum_{j=1}^{k-1} \beta_{kj}\varphi_j, \tag{3.6}$$

where

$$\beta_{kj} = \frac{\langle F_k, \varphi_j \rangle}{\langle \varphi_j, \varphi_j \rangle}, \quad k = 2, 3, \ldots, n, \; j = 1, 2, \ldots, k-1,$$

and $\langle \, , \, \rangle$ denotes inner product of two functions.

We define inner product of two functions, say, φ_i and φ_k as

$$\langle \varphi_i, \varphi_k \rangle = \int_0^1 \varphi_i(X)\varphi_k(X)\,dX, \quad i = 1, 2, \ldots, n. \tag{3.7}$$

Here we have taken weight function as 1.

Similarly, the norm of the function φ_k is defined as

$$\|\varphi_k\| = \langle \varphi_k, \varphi_k \rangle^{1/2}.$$

As such, normalized functions $\hat{\varphi_k}$ may be obtained by using

$$\hat{\varphi_k} = \frac{\varphi_k}{\|\varphi_k\|}.$$

In the bending problem in this chapter, we have considered orthonormal polynomials ($\hat{\varphi_k}$) in Eq. (3.5). As such, substituting Eq. (3.5) in Eq. (3.4) and minimizing U_T as a function of c_j's, one may obtain the following system of linear equation:

$$\sum_{j=1}^{n} a_{ij}c_j = P_c b_i, \tag{3.8}$$

where a_{ij}, b_i, and P_c are defined in this chapter.

3.1.1.2 *Timoshenko beam theory (TBT)*

In this beam theory, the strain energy may be expressed as (Wang *et al.* 2007)

$$U = \frac{1}{2}\int_0^L \left(M\frac{d\phi}{dx} + Q\left(\phi + \frac{dw}{dx}\right)\right)dx, \tag{3.9}$$

where the bending moment M and shear force Q may be obtained by setting $\bar{N}$ and all time derivatives to zero, respectively, in Eqs. (1.17) and (1.18). As such,

$$M = EI\frac{d\phi}{dx} - \mu q, \tag{3.10}$$

$$Q = k_s GA\left(\phi + \frac{dw}{dx}\right) - \mu\frac{dq}{dx}. \tag{3.11}$$

The potential energy of the transverse force q may be given as (Ghannadpour *et al.* 2013)

$$V = -\frac{1}{2}\int_0^L qw\,dx. \tag{3.12}$$

As such, the total energy U_T of the system is then written as

$$U_T = \frac{1}{2}\int_0^L \left(EI\left(\frac{d\phi}{dx}\right)^2 - \mu q\frac{d\phi}{dx} + k_s GA\left(\phi + \frac{dw}{dx}\right)^2 - \mu\frac{dq}{dx}\left(\phi + \frac{dw}{dx}\right) - qw\right)dx. \tag{3.13}$$

For applying Rayleigh–Ritz method, each of the unknown functions w and ϕ may now be expressed as the sum of series of polynomials, viz.

$$w(X) = \sum_{k=1}^{n} c_k\varphi_k, \tag{3.14}$$

$$\phi(X) = \sum_{k=1}^{n} d_k\psi_k, \tag{3.15}$$

where n is the number of terms needed in the series and c_k's and d_k's are unknowns. Here φ_k and ψ_k are polynomials which are consisting

Table 3.1 Boundary functions used for different edge conditions (TBT)

Boundary condition	f_w	f_ϕ
S–S	$X(1-X)$	1
C–S	$X(1-X)$	X
C–C	$X(1-X)$	$X(1-X)$
C–F	X	X

of boundary polynomial specifying support conditions (essential boundary conditions) multiplied by one-dimensional simple polynomials, viz. $\varphi_k = f_w X^{k-1}$ and $\psi_k = f_\phi X^{k-1}$. It may be noted that f_w and f_ϕ are the boundary functions corresponding to unknown functions w and ϕ, respectively, which are given in Table 3.1 for some of the boundary conditions. Similarly, one may find boundary functions for other boundary conditions also.

If instead of simple polynomials, Chebyshev polynomials (T_k) are used, then $\varphi_k = f_w T_k$ and $\psi_k = f_\phi T_k$.

When φ_k and ψ_k are orthogonal polynomials, then Gram–Schmidt process may similarly be used as discussed in Section 2.1.1.1 in order to obtain these polynomials from the set of linearly independent functions F_k and G_k, where $F_k = f_w f_k$ and $G_k = f_\phi f_k$ with $f_k = X^{k-1}, k = 1, 2, \ldots, n$. Then orthonormal polynomials $\hat{\varphi}_k$ may be obtained as in Section 2.1.1.1 from orthogonal polynomials φ_k. Similarly, $\hat{\psi}_k$ may also be obtained.

In the bending analysis of Timoshenko nanobeams in this chapter, we have used $\hat{\varphi}_k$ and $\hat{\psi}_k$ in Eqs. (3.14) and (3.15), respectively. As such, substituting Eqs. (3.14) and (3.15) in Eq. (3.13) and then minimizing U_T as a function of constants, one may find the following system of linear equations for TBT:

$$[K]\{Z\} = P_c\{B\}, \tag{3.16}$$

where Z is a column vector of constants, $[K]$, $\{B\}$, and P_c are defined in this chapter.

3.1.2 *Vibration problems*

We have analyzed vibration of nanobeams based on EBT and TBT. In this problem, we have considered one-dimensional simple polynomials and boundary characteristic orthogonal polynomials as shape functions in the Rayleigh–Ritz method. As such, we present below the procedure for handling the above problems.

3.1.2.1 *Euler–Bernoulli beam theory (EBT)*

For free vibration, we assume displacement function as harmonic type, viz.

$$w(x,t) = w_0(x)\sin \omega t, \tag{3.17}$$

where w_0 is the amplitude of the displacement component of free vibration of nanobeam and ω denotes natural frequency of vibration.

As such, maximum strain energy $U_{\max}$ may be given as

$$U_{\max} = -\frac{1}{2}\int_0^L M\frac{d^2 w_0}{dx^2}dx, \tag{3.18}$$

where M may be obtained from Eq. (1.8) by setting $\bar{N}$ and q to zero. As such, M may be given by

$$M = -EI\frac{d^2 w_0}{dx^2} + \mu(-\rho A\omega^2 w_0). \tag{3.19}$$

Here we have taken $m_0 = \rho A$.

Maximum kinetic energy is expressed as

$$T_{\max} = \frac{1}{2}\int_0^L \rho A\omega^2 w_0^2\, dx, \tag{3.20}$$

where ρ denotes the density of beams.

Rayleigh quotient λ^2 may be obtained by equating maximum kinetic and strain energies which is given in Chapter 5.

In the vibration of Euler–Bernoulli nanobeams (Chapter 5), simple polynomials (φ_k) and orthonormal polynomials ($\hat{\varphi}_k$) are used in Eq. (3.5). As such, substituting Eq. (3.5) in the Rayleigh quotient and minimizing Rayleigh quotient with respect to the constant coefficients, a generalized eigenvalue problem will be obtained as

$$[K]\{Z\} = \lambda^2\,[M_a]\{Z\}, \tag{3.21}$$

where Z is a column vector of constants and the matrices $[K]$ as well as $[M_a]$ are the stiffness and mass matrices which are defined in respective chapters.

3.1.2.2 *Timoshenko beam theory (TBT)*

Here also, we have considered free harmonic motion, viz. $w(x,t) = w_0(x)\sin\omega t$ and $\phi(x,t) = \phi_0(x)\sin\omega t$.

Maximum strain energy may be given as (Wang *et al.* 2007)

$$U_{\max} = \frac{1}{2}\int_0^L \left(M\frac{d\phi_0}{dx} + Q\left(\phi_0 + \frac{dw_0}{dx}\right)\right)dx. \tag{3.22}$$

Bending moment M and shear force Q may be obtained by setting $\bar{N}$ and q to zero, respectively, in Eqs. (1.17) and (1.18). As such, we have

$$M = EI\frac{d\phi_0}{dx} + \mu\left[-\rho A\omega^2 w_0 - \rho I\omega^2\frac{d\phi_0}{dx}\right], \tag{3.23}$$

$$Q = GAK_s\left(\phi_0 + \frac{dw_0}{dx}\right) - \mu\rho A\omega^2\frac{dw_0}{dx}. \tag{3.24}$$

Maximum kinetic energy is written as (Wang *et al.* 2007)

$$T_{\max} = \frac{1}{2}\int_0^L (\rho A\omega^2 w_0^2 + \rho I\omega^2\phi_0^2)\,\mathrm{d}x. \tag{3.25}$$

Equating maximum kinetic and strain energies, one may obtain Rayleigh quotient (λ^2) which is given in Chapter 5.

In the vibration of Timoshenko nanobeams (Chapter 5), we have used simple polynomials (φ_k and ψ_k) and orthonormal polynomials ($\hat{\varphi}_k$ and $\hat{\psi}_k$) in Eqs. (3.14) and (3.15).

Substituting Eqs. (3.14) and (3.15) in the Rayleigh quotient and minimizing λ^2 with respect to the unknown coefficients c_j and d_j, the following generalized eigenvalue problem will be obtained:

$$[K]\{Z\} = \lambda^2[M_a]\{Z\}, \tag{3.26}$$

where Z is a column vector of constants and the matrices $[K]$ as well as $[M_a]$ are stiffness and mass matrices which are given in Chapter 5.

3.2 Plate Theory

Here we have discussed the procedure for applying Rayleigh–Ritz method in the bending, buckling and vibration analyses of nanoplates based on CPT. Consider a rectangular nanoplate with the domain $a \le x \le b$, $a \le y \le b$ in xy-plane, where a and b are the length and the breadth of the nanoplate,

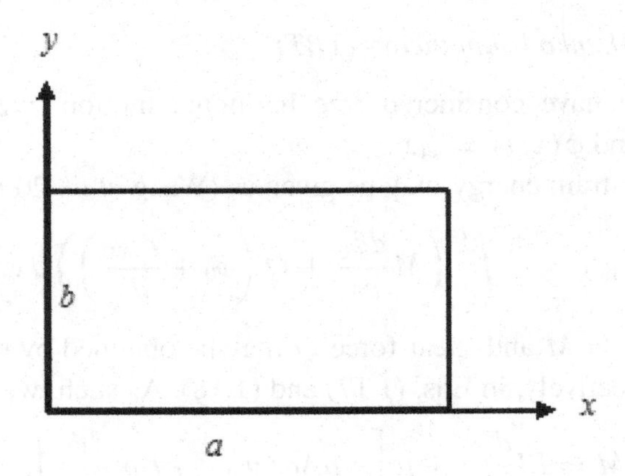

Fig. 3.1 Geometry of the nanoplate

respectively. The x- and y-axes are taken along the edges of the nanoplate and z-axis is perpendicular to the xy-plane. The middle surface being $z = 0$ and origin is at one of the corners of the nanoplate (Fig. 3.1).

3.2.1 *Classical plate theory (CPT)*

3.2.1.1 *Bending problem*

To apply Rayleigh–Ritz method, one should have knowledge about energies of the system. As such, we have shown below energies of orthotropic nanoplates embedded in elastic foundations such as Winkler and Pasternak.

The strain energy (U) of orthotropic nanoplates may be given by (Anjomshoa 2013)

$$U = \frac{1}{2}\int_0^a \int_0^b \left\{ D_{11}\left(\frac{\partial^2 w}{\partial x^2}\right)^2 + 2D_{12}\left(\frac{\partial^2 w}{\partial x^2}\frac{\partial^2 w}{\partial y^2}\right) + D_{22}\left(\frac{\partial^2 w}{\partial y^2}\right)^2 \right.$$
$$+ 4D_{33}\left(\frac{\partial^2 w}{\partial x \partial y}\right)^2 + k_w\left[w^2 + \mu\left(\left(\frac{\partial w}{\partial x}\right)^2 + \left(\frac{\partial w}{\partial y}\right)^2\right)\right]$$
$$+ k_p\left[\left(\frac{\partial w}{\partial x}\right)^2 + \left(\frac{\partial w}{\partial y}\right)^2 + \mu\left(\left(\frac{\partial^2 w}{\partial x^2}\right)^2\right.\right.$$
$$\left.\left.\left. + 2\left(\frac{\partial^2 w}{\partial x \partial y}\right)^2 + \left(\frac{\partial^2 w}{\partial y^2}\right)^2\right)\right]\right\} dx\, dy, \tag{3.27}$$

where w is the displacement and $\mu = (e_0 l_{\text{int}})^2$ is the nonlocal parameter with e_0 as material constant and l_{int} as internal characteristic length of the system (lattice parameter, granular distance, distance between C–C bonds). Here k_w and k_p denote the Winkler and Pasternak coefficients of the elastic foundation, respectively, and D_{ij} are flexural rigidities of the nanoplate which are defined as

$$D_{11} = \frac{E_x h^3}{12(1-\nu_x\nu_y)}, \quad D_{12} = \frac{\nu_y E_x h^3}{12(1-\nu_x\nu_y)} = \frac{\nu_x E_y h^3}{12(1-\nu_x\nu_y)},$$

$$D_{22} = \frac{E_y h^3}{12(1-\nu_x\nu_y)}, \quad \text{and} \quad D_{33} = \frac{G_{xy} h^3}{12}.$$

In the expression of flexural rigidities, h is the height of the nanoplate, E_x and E_y Young's moduli, ν_x and ν_y Poisson's ratios, and G_{xy} the shear modulus of the nanoplate.

In the case of isotropic nanoplates, the potential energy (Eq. (3.27)) reduces to

$$\begin{aligned} U = \frac{1}{2} D \int_0^a \int_0^b \Bigg\{ & \left(\frac{\partial^2 w}{\partial x^2}\right)^2 + 2\nu \left(\frac{\partial^2 w}{\partial x^2}\frac{\partial^2 w}{\partial y^2}\right) + \left(\frac{\partial^2 w}{\partial y^2}\right)^2 \\ & + 2(1-\nu)\left(\frac{\partial^2 w}{\partial x \partial y}\right)^2 + k_w \left[w^2 + \mu \left(\left(\frac{\partial w}{\partial x}\right)^2 + \left(\frac{\partial w}{\partial y}\right)^2 \right) \right] \\ & + k_p \Bigg[\left(\frac{\partial w}{\partial x}\right)^2 + \left(\frac{\partial w}{\partial y}\right)^2 + \mu \Bigg(\left(\frac{\partial^2 w}{\partial x^2}\right)^2 \\ & + 2\left(\frac{\partial^2 w}{\partial x \partial y}\right)^2 + \left(\frac{\partial^2 w}{\partial y^2}\right)^2 \Bigg) \Bigg] \Bigg\} \, dx \, dy, \end{aligned} \tag{3.28}$$

where $D = Eh^3/12(1-\nu^2)$ is the flexural rigidity for isotropic nanoplate.

The potential energy of the transverse force q may be given by (Anjomshoa 2013)

$$V = -q \left[w - \mu \left(\frac{\partial^2 w}{\partial x^2} + \frac{\partial^2 w}{\partial y^2} \right) \right]. \tag{3.29}$$

The total energy U_T of the system may be written as

$$U_T = U + V. \tag{3.30}$$

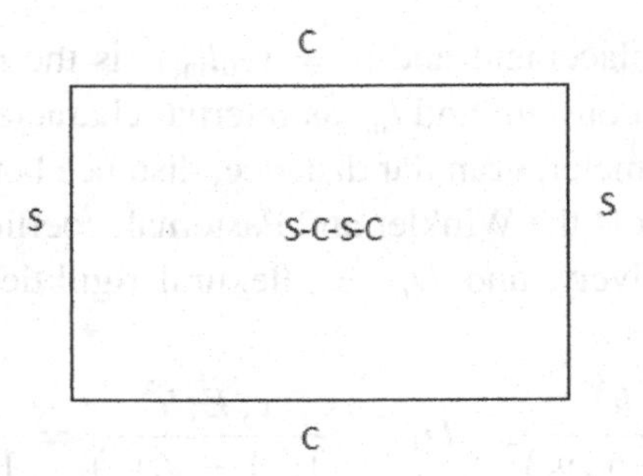

Fig. 3.2 Boundary condition

Displacement function w may be expressed as the sum of series of polynomials. As such,

$$w(X) = \sum_{k=1}^{n} c_k \varphi_k, \tag{3.31}$$

where n is the number of terms needed in the series and c_k's are unknowns. Here φ_k are the polynomials which are consisting of a boundary polynomial specifying support conditions (essential boundary conditions) multiplied by two-dimensional simple polynomials, viz. $\varphi_k = ff_k$, where f_k are two-dimensional simple polynomials and $f = X^u(1-X)^v Y^{u_1}(1-Y)^{v_1}$. Here $u = 0$, 1 or 2 as the edge $X = 0$ is free, simply supported or clamped. Same justification can be given to v, u_1, and v_1 for the edges $X = 1$, $Y = 0$, and $Y = 1$. The edge conditions are taken in anticlockwise direction starting at the edge $X = 0$ (Fig. 3.2) and obtained by assigning various values to u, v, u_1, and v_1 as 0, 1, 2 for free, simply supported, and clamped edge conditions, respectively.

When φ_k are orthogonal polynomials, then Gram–Schmidt process as discussed in Section 2.1.1.1 has been used to obtain these polynomials, respectively, from a set of linearly independent functions F_k, where $F_k = ff_k$. Then orthonormal polynomials $\hat{\varphi}_k$ may be obtained from orthogonal polynomials φ_k as discussed in Section 2.1.1.1.

It is noted here that two-dimensional simple algebraic polynomials such as $1, X, Y, X^2, XY, Y^2, X^3, X^2Y, XY^2, Y^3$, and so on have been used.

Substituting Eq. (3.31) in Eq. (3.30) and then minimizing total energy of the system as a function of constants, one may obtain the following system of linear equation:

$$\sum_{j=1}^{n} a_{ij} c_j = P_c b_i, \tag{3.32}$$

where a_{ij}, P_c, and b_i are given in Chapter 4.

3.2.1.2 *Buckling problem*

For this problem, we have the strain energy same as that of Eq. (3.27).

The potential energy due to axial compressive force is written as (Anjomshoa 2013)

$$\begin{aligned} V_a = \frac{1}{2} N_{xx} \int_0^a \int_0^b & \left\{ \left(\frac{\partial w}{\partial x} \right)^2 + \mu \left(\left(\frac{\partial^2 w}{\partial x^2} \right)^2 + \left(\frac{\partial^2 w}{\partial x \partial y} \right)^2 \right) \right. \\ & + \frac{N_{yy}}{N_{xx}} \left[\left(\frac{\partial w}{\partial y} \right)^2 + \mu \left(\left(\frac{\partial^2 w}{\partial y^2} \right)^2 + \left(\frac{\partial^2 w}{\partial x \partial y} \right)^2 \right) \right] \\ & \left. 2 \frac{N_{xy}}{N_{xx}} \left[\frac{\partial w}{\partial x} \frac{\partial w}{\partial y} + \mu \left(\frac{\partial^2 w}{\partial x^2} \frac{\partial^2 w}{\partial x \partial y} + \frac{\partial^2 w}{\partial y^2} \frac{\partial^2 w}{\partial x \partial y} \right) \right] \right\} dx\, dy. \end{aligned} \tag{3.33}$$

For uniform in-plane compression, we have used the relations $N_{xx} = N_{yy} = -N$, $N_{xy} = 0$.

Accordingly, Eq. (3.33) reduces to

$$\begin{aligned} V_a = \frac{1}{2} N_{xx} \int_0^a \int_0^b & \left\{ \left(\frac{\partial w}{\partial x} \right)^2 + \mu \left(\left(\frac{\partial^2 w}{\partial x^2} \right)^2 + \left(\frac{\partial^2 w}{\partial x \partial y} \right)^2 \right) \right. \\ & \left. + \left(\frac{\partial w}{\partial y} \right)^2 + \mu \left(\left(\frac{\partial^2 w}{\partial y^2} \right)^2 + \left(\frac{\partial^2 w}{\partial x \partial y} \right)^2 \right) \right\} dx\, dy. \end{aligned}$$

Equating above energies of the system, one may get Rayleigh quotient ($\bar{N}^0$) which is given in Chapter 4.

Substituting Eq. (3.31) into Rayleigh quotient, we get a generalized eigenvalue problem as

$$[K]\{Z\} = \bar{N}^0 [B_c]\{Z\}, \tag{3.34}$$

where K and B_c are the stiffness and buckling matrices given in Chapter 4 and Z is a column vector of constants.

3.2.1.3 *Vibration problem*

In this case, we have considered free harmonic motion.

As such, maximum strain energy for isotropic nanoplate is given by

$$U_{\max} = \frac{1}{2} D \int_0^a \int_0^b \Bigg\{ \left(\frac{\partial^2 w_0}{\partial x^2} \right)^2 + 2\nu \left(\frac{\partial^2 w_0}{\partial x^2} \frac{\partial^2 w_0}{\partial y^2} \right) + \left(\frac{\partial^2 w_0}{\partial y^2} \right)^2$$
$$+2(1-\nu) \left(\frac{\partial^2 w_0}{\partial x \partial y} \right)^2 + k_w \left[w_0^2 + \mu \left(\left(\frac{\partial w_0}{\partial x} \right)^2 + \left(\frac{\partial w_0}{\partial y} \right)^2 \right) \right]$$
$$+k_p \left[\left(\frac{\partial w_0}{\partial x} \right)^2 + \left(\frac{\partial w_0}{\partial y} \right)^2 + \mu \left(\left(\frac{\partial^2 w_0}{\partial x^2} \right)^2 \right. \right.$$
$$\left. \left. +2 \left(\frac{\partial^2 w_0}{\partial x \partial y} \right)^2 + \left(\frac{\partial^2 w_0}{\partial y^2} \right)^2 \right) \right] \Bigg\} dx\, dy. \tag{3.35}$$

Maximum kinetic energy ($T_{\max}$) is given by (Anjomshoa 2013)

$$T_{\max} = \frac{1}{2} m_0 \omega^2 \int_0^a \int_0^b \left\{ w_0^2 + \mu \left(\left(\frac{\partial w_0}{\partial x} \right)^2 + \left(\frac{\partial w_0}{\partial y} \right)^2 \right) \right\} dx\, dy. \tag{3.36}$$

Equating maximum kinetic and strain energies, one may obtain the Rayleigh quotient (λ^2).

Substituting Eq. (3.31) into Rayleigh quotient (λ^2), we get a generalized eigenvalue problem as

$$[K]\{Z\} = \lambda^2 [M_a]\{Z\}, \tag{3.37}$$

where K and M_a are, respectively, the stiffness and mass matrices which are given in Chapter 7.

3.3 Differential Quadrature Method (DQM)

We have used DQM in the buckling and vibration analyses of nanobeams. As such, we have first shown governing equations which are then converted into single variable.

3.3.1 *Buckling problems*

We have shown below governing differential equations for buckling analyses of nanobeams based on four types of beam theories such as EBT, TBT, RBT, and LBT.

3.3.2 *Euler–Bernoulli beam theory (EBT)*

Following governing equation for buckling analysis may be obtained from Eq. (1.9) by setting q and all time derivatives as zero:

$$-EI\frac{d^4w}{dx^4} + \mu\bar{N}\frac{d^4w}{dx^4} - \bar{N}\frac{d^2w}{dx^2} = 0. \tag{3.38}$$

3.3.3 *Timoshenko beam theory (TBT)*

For this theory, the governing equation is obtained from Eqs. (1.19) and (1.20) by setting q and all time derivatives as zero and is written as

$$GAK_s\left(\frac{d\phi}{dx} + \frac{d^2w}{dx^2}\right) - \bar{N}\frac{d^2w}{dx^2} + \mu\bar{N}\frac{d^4w}{dx^4} = 0, \tag{3.39}$$

$$EI\frac{d^2\phi}{dx^2} - GAK_s\left(\phi + \frac{dw}{dx}\right) = 0. \tag{3.40}$$

Eliminating ϕ from Eqs. (3.39) and (3.40), governing equations can be written in terms of one variable as

$$\begin{aligned} -EI\frac{d^4w}{dx^4} + EI\frac{\bar{N}}{k_sGA}\frac{d^4w}{dx^4} - \mu EI\frac{\bar{N}}{k_sGA}\frac{d^6w}{dx^6} \\ -\bar{N}\frac{d^2w}{dx^2} + \mu\bar{N}\frac{d^4w}{dx^4} = 0. \end{aligned} \tag{3.41}$$

3.3.4 *Reddy–Bickford beam theory (RBT)*

The governing equation for RBT may be obtained from Eqs. (1.26) and (1.27) by setting q and all time derivatives as zero:

$$G\tilde{A}\left(\frac{d\phi}{dx} + \frac{d^2w}{dx^2}\right) - \bar{N}\frac{d^2w}{dx^2} + \mu\bar{N}\frac{d^4w}{dx^4}$$

$$+c_1 EJ\frac{d^3\phi}{dx^3} - c_1^2 EK\left(\frac{d^3\phi}{dx^3} + \frac{d^4 w}{dx^4}\right) = 0, \tag{3.42}$$

$$E\hat{I}\frac{d^2\phi}{dx^2} - c_1 E\hat{J}\left(\frac{d^2\phi}{dx^2} + \frac{d^3 w}{dx^3}\right) - G\bar{A}\left(\phi + \frac{dw}{dx}\right) = 0. \tag{3.43}$$

Again eliminating ϕ from Eqs. (3.42) and (3.43), governing equation is obtained in terms of displacement as

$$\begin{aligned}\frac{105}{84}G\tilde{A}\frac{d^4 w}{dx^4} - \frac{1}{105}EI\frac{d^6 w}{dx^6} &= -\frac{105}{84EI}G\tilde{A}\bar{N}\frac{d^2 w}{dx^2} + \frac{105}{84EI}\mu G\tilde{A}\bar{N}\frac{d^4 w}{dx^4}\\ &\quad + \frac{68}{84}\bar{N}\frac{d^4 w}{dx^4} - \frac{68}{84}\mu\bar{N}\frac{d^6 w}{dx^6}.\end{aligned} \tag{3.44}$$

3.3.5 *Levinson beam theory (LBT)*

In this case, the governing equation is obtained from Eqs. (1.33) and (1.34) by again setting q and all time derivatives as zero:

$$G\bar{A}\left(\frac{d\phi}{dx} + \frac{d^2 w}{dx^2}\right) - \bar{N}\frac{d^2 w}{dx^2} + \mu\bar{N}\frac{d^4 w}{dx^4} = 0, \tag{3.45}$$

$$EI\frac{d^2\phi}{dx^2} - c_1 EJ\left(\frac{d^2\phi}{dx^2} + \frac{d^3 w}{dx^3}\right) - G\bar{A}\left(\phi + \frac{dw}{dx}\right) = 0. \tag{3.46}$$

Eliminating ϕ from Eqs. (3.45) and (3.46), the reduced governing differential equation is

$$\frac{4}{5}EI\frac{\bar{N}}{G\bar{A}}\frac{d^4 w}{dx^4} - EI\frac{d^4 w}{dx^4} - \frac{4}{5}EI\mu\frac{\bar{N}}{G\bar{A}}\frac{d^6 w}{dx^6} - \bar{N}\frac{d^2 w}{dx^2} + \mu\bar{N}\frac{d^4 w}{dx^4} = 0. \tag{3.47}$$

3.3.6 *Vibration problems*

Here, we have shown governing equations for free vibration analysis based on four beam theories such as EBT, TBT, RBT, and LBT. Here free harmonic motion has been assumed. Also we have used the relations $m_0 = \rho A$ and $m_2 = \rho I$.

3.3.7 *Euler–Bernoulli beam theory (EBT)*

For EBT, the governing equation for vibration analysis may be obtained from Eq. (1.9) by setting q and $\bar{N}$ to zero and is written as

$$EI\frac{d^4w_0}{dx^4} + \mu\rho A\omega^2\frac{d^2w_0}{dx^2} = \rho A\omega^2 w_0. \tag{3.48}$$

3.3.8 *Timoshenko beam theory (TBT)*

In this case, the governing equation may be obtained from Eqs. (1.19) and (1.20) by setting q and $\bar{N}$ to zero and we have

$$GAK_s\left(\frac{d\phi_0}{dx} + \frac{d^2w_0}{dx^2}\right) = \rho A\omega^2\left(-w_0 + \mu\frac{d^2w_0}{dx^2}\right), \tag{3.49}$$

$$EI\frac{d^2\phi_0}{dx^2} - GAK_s\left(\phi_0 + \frac{dw_0}{dx}\right) = \rho I\omega^2\left(-\phi_0 + \mu\frac{d^2\phi_0}{dx^2}\right). \tag{3.50}$$

Eliminating ϕ_0 from Eqs. (3.49) and (3.50) and neglecting coefficient of ω_n^4, the governing equations can be transformed to

$$\begin{aligned} -EI\frac{d^4w_0}{dx^4} &= \frac{EI}{k_sGA}\rho A\omega^2\left(\frac{d^2w_0}{dx^2} - \mu\frac{d^4w_0}{dx^4}\right) \\ &\quad + \rho I\omega^2\left(\frac{d^2w_0}{dx^2} - \mu\frac{d^4w_0}{dx^4}\right) - \rho A\omega^2\left(w_0 - \mu\frac{d^2w_0}{dx^2}\right). \end{aligned} \tag{3.51}$$

3.3.9 *Reddy–Bickford beam theory (RBT)*

The governing equation for RBT may be obtained from Eqs. (1.26) and (1.27) by setting q and $\bar{N}$ to zero and is obtained as

$$\begin{aligned} &G\tilde{A}\left(\frac{d\phi_0}{dx} + \frac{d^2w_0}{dx^2}\right) + c_1EJ\frac{d^3\phi_0}{dx^3} - c_1^2EK\left(\frac{d^3\phi_0}{dx^3} + \frac{d^4w_0}{dx^4}\right) \\ &= -\rho A\omega^2\left(w_0 - \mu\frac{d^2w_0}{dx^2}\right), \end{aligned} \tag{3.52}$$

$$E\hat{I}\frac{d^2\phi_0}{dx^2} - c_1E\hat{J}\left(\frac{d^2\phi_0}{dx^2} + \frac{d^3w_0}{dx^3}\right) - G\tilde{A}\left(\phi_0 + \frac{dw_0}{dx}\right) = 0. \tag{3.53}$$

Eliminating ϕ_0 from Eqs. (3.52) and (3.53), the governing equations may be obtained in terms of w_0 as

$$G\tilde{A}\frac{5}{4}\frac{d^4 w_0}{dx^4} - \frac{1}{105}EI\frac{d^6 w_0}{dx^6}$$
$$= -\frac{17}{21}\rho A\omega^2\frac{d^2 w_0}{dx^2} + \frac{17}{21}\mu\rho A\omega^2\frac{d^4 w_0}{dx^4}$$
$$+ G\tilde{A}\frac{105}{84EI}\rho A\omega^2\left(w_0 - \mu\frac{d^2 w_0}{dx^2}\right). \tag{3.54}$$

3.3.10 *Levinson beam theory (LBT)*

For this beam theory, the governing equations may be obtained from Eqs. (1.33) and (1.34) by setting q and $\bar{N}$ to zero and are written as

$$G\bar{A}\left(\frac{d\phi_0}{dx} + \frac{d^2 w_0}{dx^2}\right) = -\rho A\omega^2\left(w_0 - \mu\frac{d^2 w_0}{dx^2}\right), \tag{3.55}$$

$$EI\frac{d^2\phi_0}{dx^2} - c_1 EJ\left(\frac{d^2\phi_0}{dx^2} + \frac{d^3 w_0}{dx^3}\right) - G\bar{A}\left(\phi_0 + \frac{dw_0}{dx}\right)$$
$$= -\rho I\omega^2\left(\phi_0 - \mu\frac{d^2\phi_0}{dx^2}\right). \tag{3.56}$$

Eliminating ϕ_0 from Eqs. (3.55) and (3.56) and neglecting coefficient of ω_n^4, the governing differential equation reduces to

$$\left(-\rho A\omega^2\frac{d^2 w_0}{dx^2} + \rho A\mu\omega^2\frac{d^4 w_0}{dx^4} - G\bar{A}\frac{d^4 w_0}{dx^4}\right)\left(\frac{EI}{G\bar{A}} - \frac{c_1 EJ}{G\bar{A}}\right)$$
$$- c_1 EJ\frac{d^4 w_0}{dx^4} + \rho A\omega^2\left(w_0 - \mu\frac{d^2 w_0}{dx^2}\right)$$
$$= \rho I\omega^2\left(\frac{d^2 w_0}{dx^2} - \mu\frac{d^4 w_0}{dx^4}\right). \tag{3.57}$$

Next, we have briefly explained the procedure for applying DQM in the above equations. We have considered the function $w(X)$ in the domain [0, 1] with n discrete grid points. Then the first derivative at point i, $w_i{}' = dw/dX$ at $X = X_i$ ($X_1 = 0$ and $X_n = 0$) is given by (Wang and Bert 1993)

$$\begin{aligned} w_i{}' &= \sum_{j=1}^{n} A_{ij} w_j, \\ w_i{}'' &= \sum_{j=1}^{n} B_{ij} w_j, \\ w_i{}''' &= \sum_{j=1}^{n} C_{ij} w_j, \\ w_i{}^{IV} &= \sum_{j=1}^{n} D_{ij} w_j, \end{aligned} \tag{3.58}$$

where $i = 1, 2, \ldots, n$ and n is the number of discrete grid points.

Here A_{ij}, B_{ij}, C_{ij}, and D_{ij} are the weighting coefficients of the first, second, third, and fourth derivatives, respectively.

3.3.10.1 *Determination of weighting coefficients*

Computation of weighting coefficient matrix $A = (A_{ij})$ is the key step in the DQM. In the present investigation, we have used Quan and Chang's (1989) approach to compute weighting coefficients A_{ij}. As per this approach, matrix $A = (A_{ij})$ may be computed by the following procedure:

$$A_{ij} = \frac{1}{X_j - X_i} \prod_{\substack{k \neq i \\ k \neq j \\ k=1}}^{n} \frac{X_i - X_k}{X_j - X_k}, \quad i \neq j,$$

$$i = 1, 2, \ldots, n, \quad j = 1, 2, \ldots, n, \tag{3.59}$$

$$A_{ii} = \sum_{\substack{k \neq i \\ k=1}}^{n} \frac{1}{X_i - X_k}, \quad i = j, \quad i = 1, 2, \ldots, n. \tag{3.60}$$

Once weighting coefficients of first-order derivatives are computed, weighting coefficients of higher-order derivatives may be obtained by simply matrix multiplication as follows:

$$B = B_{ij} = \sum_{k=1}^{n} A_{ik} A_{kj}, \tag{3.61}$$

$$C = C_{ij} = \sum_{k=1}^{n} A_{ik} B_{kj}, \tag{3.62}$$

$$D = D_{ij} = \sum_{k=1}^{n} A_{ik} C_{kj} = \sum_{k=1}^{n} B_{ik} B_{kj}. \tag{3.63}$$

3.3.10.2 *Selection of mesh point distribution*

We assume that the domain $0 \leq X \leq 1$ is divided into $(n-1)$ intervals with coordinates of the grid points given as $X_1, X_2, \ldots, X_n$. These X_i's have been computed by using Chebyshev–Gauss–Lobatto grid points. That is

$$X_i = \frac{1}{2}\left[1 - \cos\left(\frac{i-1}{n-1}.\Pi\right)\right].$$

3.3.10.3 *Application of boundary condition*

Above matrices A, B, C, and D are converted into modified weighting coefficient matrices $\bar{A}$, $\bar{B}$, $\bar{C}$, and $\bar{D}$ as per the boundary condition.

First, we denote

$$\bar{A}_1 = \begin{bmatrix} 0 & A_{1,2} & \cdots & A_{1,n} \\ 0 & A_{2,2} & \cdots & A_{2,n} \\ \cdots & \cdots & \cdots & \cdots \\ 0 & A_{n,2} & \cdots & A_{n,n} \end{bmatrix},$$

$$\bar{A}_2 = \begin{bmatrix} A_{1,1} & A_{1,2} & \cdots & A_{1,n-1} & 0 \\ A_{2,1} & A_{2,2} & \cdots & A_{2,n-1} & 0 \\ \cdots & \cdots & \cdots & \cdots & \cdots \\ A_{n,1} & A_{n,2} & \cdots & A_{n-1,n-1} & 0 \end{bmatrix}.$$

In view of the above, we now illustrate the procedure for finding modified weighting coefficient matrices as per the considered boundary conditions and are discussed below.

Simply supported–simply supported (S–S)

For this boundary condition, Eq. (3.58) may be rewritten in the matrix form as

$$A = \begin{bmatrix} A_{1,1} & A_{1,2} & \cdots & A_{1,n-1} & A_{1,n} \\ A_{2,1} & A_{2,2} & \cdots & A_{2,n-1} & A_{2,n} \\ \vdots & \vdots & & \vdots & \vdots \\ A_{n,1} & A_{n,2} & \cdots & A_{n,n-1} & A_{n,n} \end{bmatrix} \begin{Bmatrix} w_1 \\ w_2 \\ \vdots \\ w_n \end{Bmatrix} = \begin{Bmatrix} w_1' \\ w_2' \\ \vdots \\ w_n' \end{Bmatrix}. \tag{3.64}$$

Firstly, to apply boundary condition $w_1 = w_n = 0$, Eq. (3.64) becomes

$$\begin{bmatrix} 0 & A_{1,2} & \cdots & A_{1,n-1} & 0 \\ 0 & A_{2,2} & \cdots & A_{2,n-1} & 0 \\ \vdots & \vdots & & \vdots & \vdots \\ 0 & A_{n,2} & \cdots & A_{n,n-1} & 0 \end{bmatrix} \begin{Bmatrix} w_1 \\ w_2 \\ \vdots \\ w_n \end{Bmatrix} = \begin{Bmatrix} w_1' \\ w_2' \\ \vdots \\ w_n' \end{Bmatrix}, \tag{3.65}$$

or

$$[\bar{A}]\{w\} = \{w'\}. \tag{3.66}$$

For the second derivative, one has

$$\{w''\} = [A]\{w'\}. \tag{3.67}$$

Using Eq. (3.66), one may obtain

$$\begin{aligned} \{w''\} &= [A][\bar{A}]\{w\} \\ &= [\bar{B}]\{w\}, \end{aligned} \tag{3.68}$$

where $\bar{B} = [A][\bar{A}]$.

Now, since $w_1'' = w_n'' = 0$, we have the third-order derivative as

$$\{w'''\} = [\bar{A}]\{w''\}. \tag{3.69}$$

Using Eq. (3.68), one obtains

$$\begin{aligned} \{w'''\} &= [\bar{A}][\bar{B}]\{w\} \\ &= [\bar{C}]\{w\}, \end{aligned} \tag{3.70}$$

where $[\bar{C}] = [\bar{A}][\bar{B}]$.

Similarly, for the fourth-order derivative, we have

$$\begin{aligned}\{w^{IV}\} &= [A]\{w'''\} \\ &= [A][\bar{C}]\{w\} \\ &= [\bar{B}][\bar{B}]\{w\} \\ &= [\bar{D}]\{w\},\end{aligned} \tag{3.71}$$

where $[\bar{D}] = [\bar{B}][\bar{B}]$ or $[\bar{D}] = [A][\bar{C}]$.

Proceeding in the similar fashion as that of simply supported–simply supported, we have following modified coefficient matrices for other boundary conditions.

Clamped–simply supported (C–S)

$$\begin{aligned}\{w'\} &= [\bar{A}]\{w\}, \\ \{w''\} &= [\bar{A}_1]\{w'\} = [\bar{A}_1][\bar{A}]\{w\} = [\bar{B}]\{w\} \quad \text{with } [\bar{B}] = [\bar{A}_1][\bar{A}], \\ \{w'''\} &= [\bar{A}_2]\{w''\} = [\bar{A}_2][\bar{B}]\{w\} = [\bar{C}]\{w\} \quad \text{with } [\bar{C}] = [\bar{A}_2][\bar{B}], \\ \{w^{IV}\} &= [A]\{w'''\} = [A][\bar{C}]\{w\} = [\bar{D}]\{w\} \quad \text{with } [\bar{D}] = [A][\bar{C}].\end{aligned}$$

Clamped–clamped (C–C)

$$\begin{aligned}\{w'\} &= [\bar{A}]\{w\}, \\ \{w''\} &= [\bar{A}]\{w'\} = [\bar{A}][\bar{A}]\{w\} = [\bar{B}]\{w\} \quad \text{with } [\bar{B}] = [\bar{A}][\bar{A}], \\ \{w'''\} &= [A]\{w''\} = [A][\bar{B}]\{w\} = [\bar{C}]\{w\} \quad \text{with } [\bar{C}] = [A][\bar{B}], \\ \{w^{IV}\} &= [A]\{w'''\} = [A][\bar{C}]\{w\} = [\bar{D}]\{w\} \quad \text{with } [\bar{D}] = [A][\bar{C}].\end{aligned}$$

Clamped–free (C–F)

$$\begin{aligned}\{w'\} &= [\bar{A}_1]\{w\}, \\ \{w''\} &= [\bar{A}_1]\{w'\} = [\bar{A}_1][\bar{A}_1]\{w\} = [\bar{B}]\{w\} \quad \text{with } [\bar{B}] = [\bar{A}_1][\bar{A}_1], \\ \{w'''\} &= [\bar{A}_2]\{w''\} = [\bar{A}_2][\bar{B}]\{w\} = [\bar{C}]\{w\} \quad \text{with } [\bar{C}] = [\bar{A}_2][\bar{B}], \\ \{w^{IV}\} &= [\bar{A}_2]\{w'''\} = [\bar{A}_2][\bar{C}]\{w\} = [\bar{D}]\{w\} \quad \text{with } [\bar{D}] = [\bar{A}_2][\bar{C}].\end{aligned}$$

It may be noted that while substituting values of the derivatives in the governing differential equations, one has to use $[\bar{A}]$, $[\bar{B}]$, $[\bar{C}]$, and $[\bar{D}]$ as per the specified boundary conditions.

Substituting Eq. (3.58) in any of the Eqs. (3.38), (3.41), (3.44), and (3.47) depending upon the beam theories, a generalized eigenvalue problem for buckling problem obtained as

$$[K]\{W\} = \bar{N}^0 [B_c]\{W\}, \tag{3.72}$$

where K is the stiffness matrix, B_c is the buckling matrix, and $\bar{N}^0$ is the buckling load parameter.

Substituting Eq. (3.58) in any of the Eqs. (3.48), (3.51), (3.54), and (3.57) depending upon the beam theories, a generalized eigenvalue problem obtained for vibration problem as

$$[K]\{W\} = \lambda^2 [M_a]\{W\}, \tag{3.73}$$

where K is the stiffness matrix, M_a is the mass matrix, and λ^2 is the frequency parameter.

Chapter 4

Bending of Nanobeams

In this chapter, bending analysis has been carried out based on Euler–Bernoulli beam theories (EBTs) and Timoshenko beam theories (TBTs) in conjunction with nonlocal elasticity theory of Eringen (1972). Boundary characteristic orthogonal polynomials have been used as shape functions in the Rayleigh–Ritz method. Various parametric studies have been carried out and shown graphically. Deflection and rotation shapes of nanobeams with specified boundary conditions have also been presented.

At first, non-dimensionalization may be done by introducing variable X as

$$X = \frac{x}{L}.$$

In this problem, we have considered a uniform transverse distributed load, viz. $q(X) = q_0$. Both EBTs and TBTs are considered. Rayleigh–Ritz method with boundary characteristic orthogonal polynomials as shape functions has been applied. Application of the method converts bending problem into system of linear equations as discussed in Section 2.1.1.1. In the system of linear equation for EBT (Eq. (3.8)), the notation a_{ij}, b_i, and P_c are defined as follows:

$$P_c = \frac{q_0 L^4}{EI},$$

$$a_{ij} = \int_0^1 \hat{\varphi}_i'' \hat{\varphi}_j'' \, dX,$$

$$b_i = \int_0^1 \left(\hat{\varphi}_i - \frac{\mu}{L^2} \hat{\varphi}_i'' \right) dX,$$

where $i = 1, 2, \ldots, n$ and $j = 1, 2, \ldots, n$.

Similarly, in the system of linear equation for TBT (Eq. (3.16)), the notation $[K]$ and $\{B\}$ are given as

$$B = \begin{Bmatrix} b_1 \\ b_2 \end{Bmatrix},$$

where

$$b_1 = \begin{Bmatrix} \int_0^1 \varphi_1 \, dX \\ \vdots \\ \int_0^1 \varphi_i \, dX \end{Bmatrix}, \quad b_2 = \begin{Bmatrix} \int_0^1 \mu q L \varphi_1' \\ \vdots \\ \int_0^1 \mu q L \varphi_1' \, dX \end{Bmatrix},$$

$$K = \begin{bmatrix} k_1 & k_2 \\ k_3 & k_4 \end{bmatrix},$$

where k_1, k_2, k_3, and k_4 are submatrices which are given by

$$k_1(i,j) = \int_0^1 2k_s GA \hat{\varphi}_i' \hat{\varphi}_j' \, dX,$$

$$k_2(i,j) = \int_0^1 2k_s GAL \hat{\varphi}_i' \hat{\psi}_j \, dX,$$

$$k_3(i,j) = \int_0^1 2k_s GAL \hat{\psi}_i \hat{\varphi}_j' \, dX,$$

$$k_4(i,j) = \int_0^1 (2k_s GAL^2 \hat{\psi}_i \hat{\psi}_j + 2EI \hat{\psi}_i' \hat{\psi}_j') dX.$$

4.1 Numerical Results and Discussions

The parameters used in this investigation are (Alshorbagy *et al.* 2013; Reddy 2007): $E = 30 \times 10^6$, $h = 1$, $k_s = \frac{5}{6}$, $nu = 0.3$. A uniformly distributed load ($q_0 = 1$) has been taken into consideration. System of linear equations have been solved by using MATLAB. It is a well-known fact that non-dimensional maximum deflection is evaluated at the center of the beam which is given by $W_{\max} = -w \times 10^2 (EI/q_0 L^4)$. At first, convergence study has been carried out to find minimum number of terms required for computation. As such, Fig. 4.1 shows convergence of EBT nanobeams for $L/h = 10$ and $\mu = 1.5\ \text{nm}^2$ with C–S support. One may note that $n = 4$ is sufficient for obtaining converged results. Next, the obtained results are compared with available literature and is shown in Table 4.1. One may find close agreement of the results.

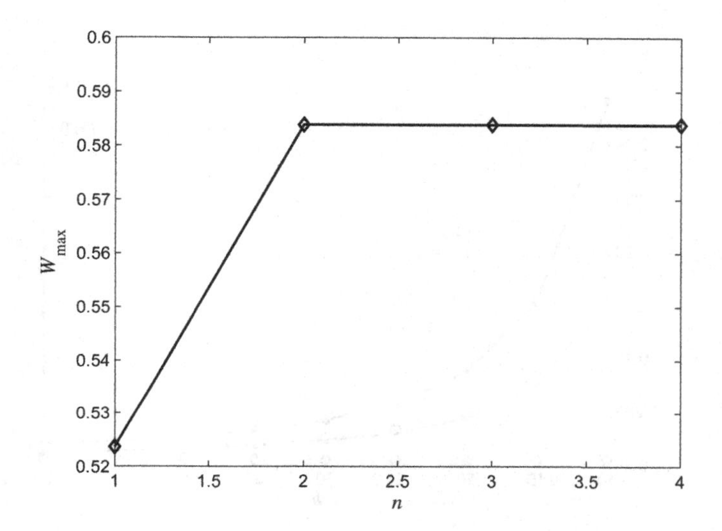

Fig. 4.1 Convergence of non-dimensional maximum center deflection for EBT

Table 4.1 Comparison of non-dimensional maximum center deflection (W_{max}) for C–S and C–C boundary conditions

	C–S		C–C	
μ	Present	Ref.*	Present	Ref.*
0	0.50	0.52	0.24	0.0.26
1	0.52	0.58	0.24	0.26
2	0.59	0.61	0.24	0.26
3	0.60	0.64	0.24	0.26

*Alshorbagy *et al.* (2013).

Next, we have carried out some of the parametric studies which are discussed below. It is noted here that unless mentioned, defection and rotation would denote non-dimensional maximum center deflection and non-dimensional maximum center rotation, respectively.

4.1.1 *Effect of aspect ratio*

Figure 4.2 illustrates the effect of aspect ratio on the deflection of nanobeams. Here, we have shown variation of deflection with an aspect

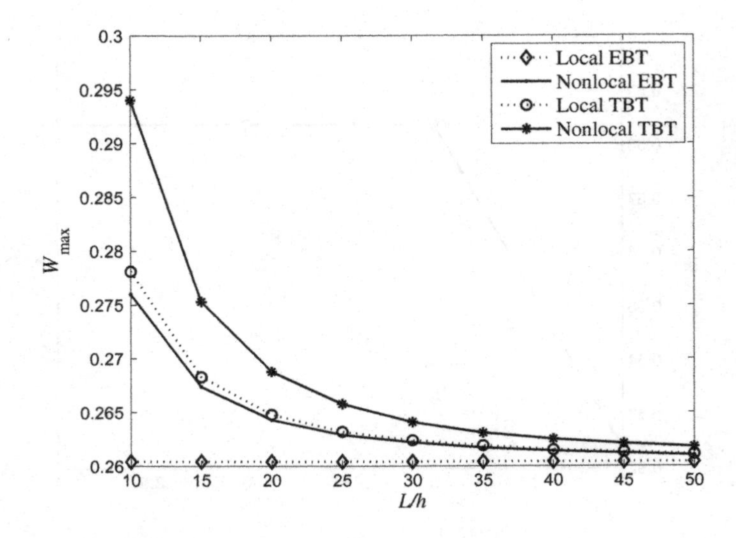

Fig. 4.2 Effect of the aspect ratio on the deflection

ratio for both local and nonlocal cases. The figure is plotted for both EBT and TBT. Nonlocal results have been computed for $\mu = 1$ nm^2. The aspect ratio (*L*/*h*) varies from 10 to 50 with the consideration of C–S boundary condition. One may observe that in the case of local EBT, the aspect ratio has no effect on the beam deflection, whereas in nonlocal EBT, deflection is dependent on the aspect ratio. It may also be noticed that in the case of both local and nonlocal TBT, the deflection is dependent on the aspect ratio. The dependency of the responses on the aspect ratio for local TBT is unique due to the effect of shear deformation. As the aspect ratio decreases, the difference between the solutions of EBT and TBT becomes highly important.

4.1.2 *Effect of scale coefficient*

Effect of scale coefficient on the bending response of nanobeams has been demonstrated in Fig. 4.3 for different values of *L*/*h*. Results have been given for TBT nanobeams with $L/h = 10$ and C–S edge condition. It is seen from the figure that bending responses vary nonlinearly with the scale coefficient. One may also observe that bending responses of nanobeams with lower aspect ratios are strongly affected by the scale coefficient than those of nanobeams with relatively higher aspect ratios. Hence, one may conclude

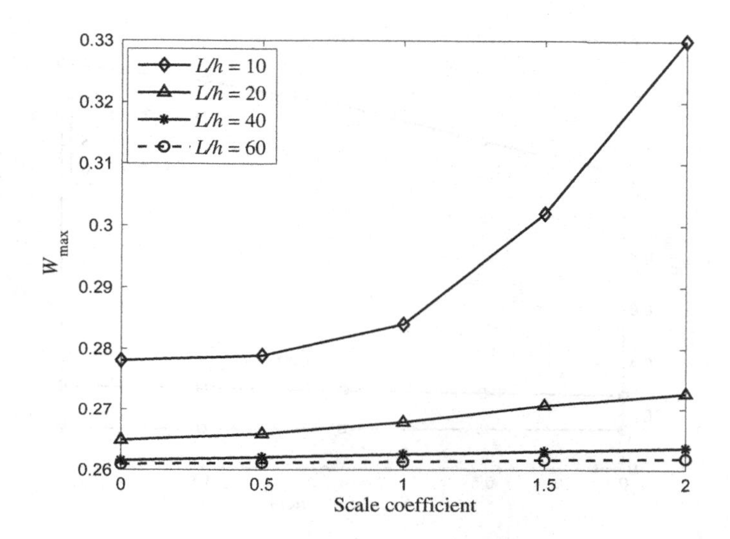

Fig. 4.3 Effect of the scale coefficient on the deflection

that local beam model may not be suitable for adequate approximation for the nanosized structures.

4.1.3 *Effect of boundary conditions*

Non-dimensional maximum center deflection of nanobeams under uniform load have been computed for different boundary conditions and have been shown graphically in Fig. 4.4. It is observed that C–C is having smallest deflection for a particular value of nonlocal parameter. One may note that in the case of C–C edge condition, there is no effect of the nonlocal parameter on the deflection, whereas in the case of S–S and C–S supports, deflection increases with increase in nonlocal parameter. Hence the effect of nonlocal parameter on the deflection is inconsistent for different boundary conditions.

4.1.4 *Deflection and rotation shapes*

In this section, we have examined the behavior of deflection and rotation shapes of nanobeams along its length for different boundary conditions. Figures 4.5–4.7 show variation of deflection with length for S–S, C–S, and C–C edge conditions, respectively. It is observed from the figures that deflection of S–S and C–S nanobeams increases with increase in nonlocal

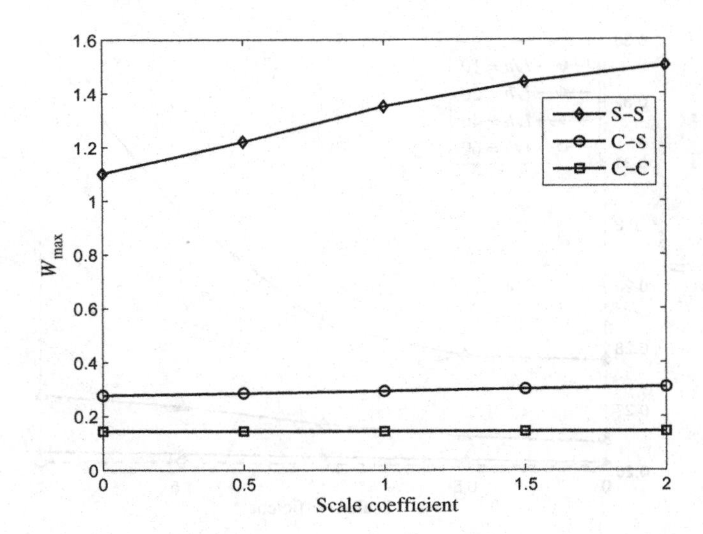

Fig. 4.4 Effect of the scale coefficient on the deflection for different boundary conditions

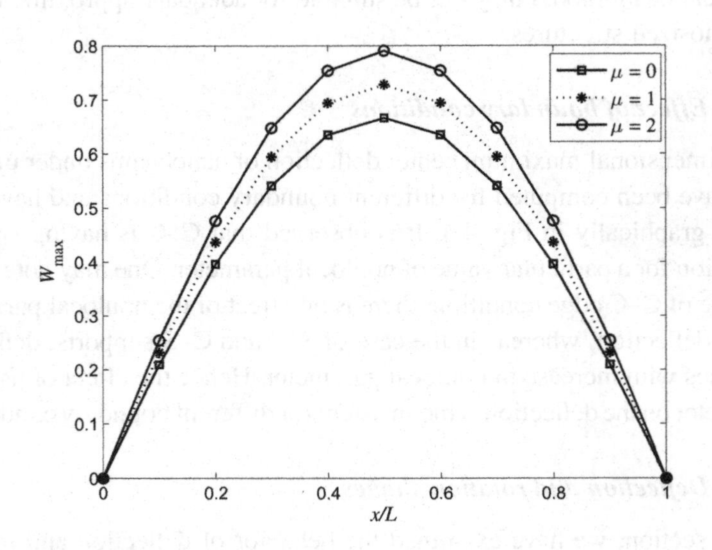

Fig. 4.5 Static deflection of S–S nanobeams for different nonlocal parameters

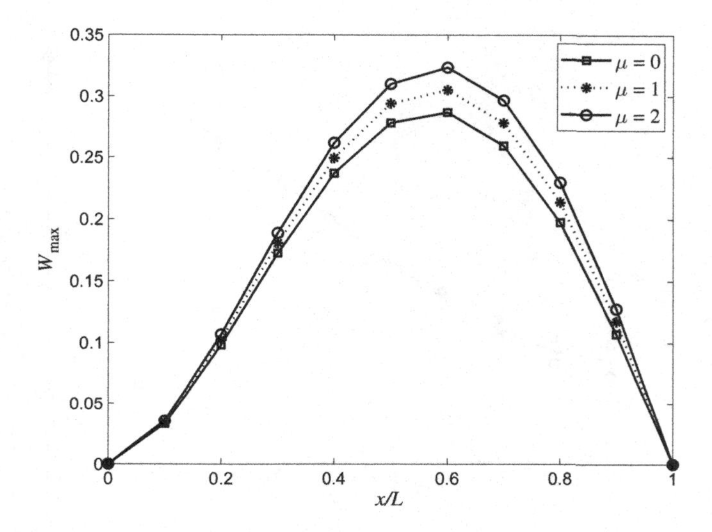

Fig. 4.6 Static deflection of C–S nanobeams for different nonlocal parameters

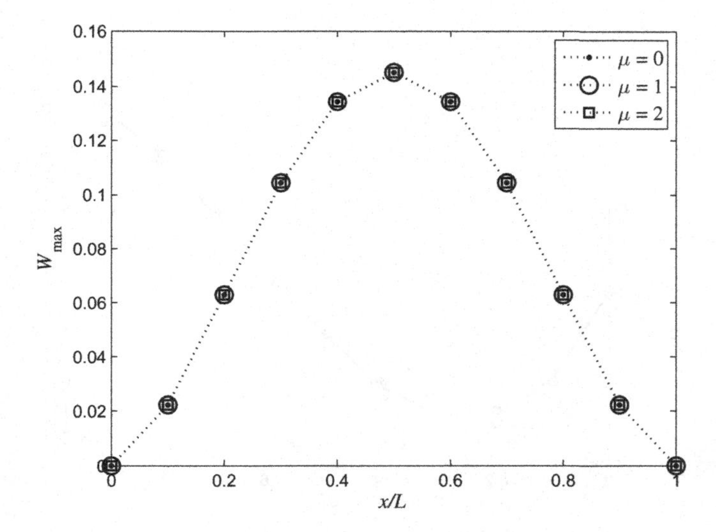

Fig. 4.7 Static deflection of C–C nanobeams for different nonlocal parameters

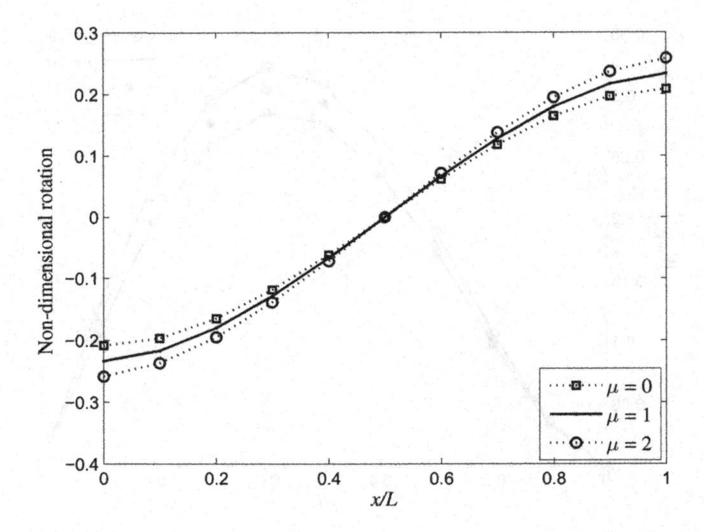

Fig. 4.8 Static rotation of S–S nanobeams for different nonlocal parameters

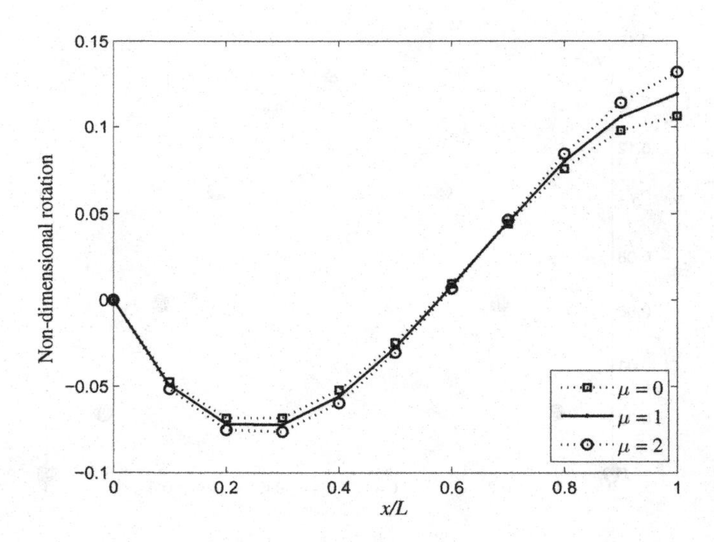

Fig. 4.9 Static rotation of C–S nanobeams for different nonlocal parameters

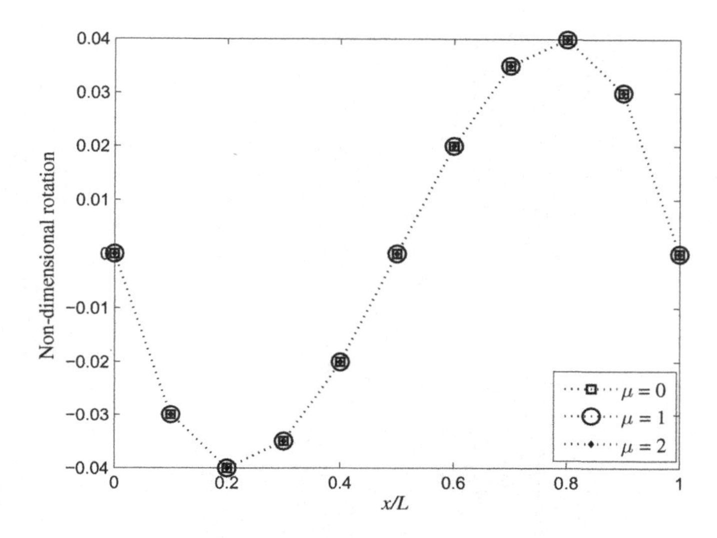

Fig. 4.10 Static rotation of C–C nanobeams for different nonlocal parameters

parameter. It is due to the fact that increasing nonlocal parameter causes increase in bonding force of atoms and this force is constraint from its boundaries which increases deflection (Alshorbagy *et al.* 2013). Another observation is seen that the nonlocal parameter has no effect on the deflection of C–C nanobeams because of its constrained nature. Next, we have shown variation of rotation with length for S–S, C–S, and C–C edge conditions, respectively, in Figs. 4.8–4.10. It may be noticed that rotation behaves differently than that of deflection. Increasing nonlocal parameter decreases rotation of S–S and C–C nanobeams up to mid-length and afterwards increase in nonlocal parameter increases rotation. One may also notice that the nonlocal parameter has no effect on the rotation of C–C nanobeams.

4.2 Conclusions

Rayleigh–Ritz method has been used for bending of nanobeams based on both EBT and TBT in conjunction with nonlocal elasticity of Eringen. The nonlocal parameter has no effect on the deflection of C–C nanobeams, whereas in the case of S–S and C–S supports, deflection increases with increase in nonlocal parameter.

Chapter 5

Buckling of Nanobeams

In this chapter, differential quadrature method (DQM) has been applied for buckling of non-uniform nanobeams based on four beam theories such as Euler–Bernoulli beam theory (EBT), Timoshenko beam theory (TBT), Reddy–Bickford beam theory (RBT), and Levinson beam theory (LBT). Here, we have also investigated buckling of nanobeams embedded in elastic medium such as Winkler and Pasternak under the influence of temperature. Boundary characteristic orthogonal polynomials and Chebyshev polynomials have been applied in the Rayleigh–Ritz method to investigate buckling of embedded nanobeams based on EBT and TBT, respectively. Also, the DQM has been employed to study buckling of embedded nanobeams based on RBT. Various parametric studies have been carried out and have been shown graphically.

As we have discussed in Chapter 1, the study of non-uniform nanobeams play a vital role for the design of nanodevices. As such, we have investigated buckling of non-uniform nanobeams having exponentially varying stiffness. Four types of beam theories have been taken into consideration. DQM has been employed and the boundary conditions are substituted in the coefficient matrices. Some of the new results in terms of boundary conditions have also been shown.

In this problem, we have assumed exponential variation of the flexural stiffness (EI) as

$$EI = EI_0 e^{-\eta X},$$

where I_0 is the second moment of area at the left end and η is the positive constant.

Here, the following non-dimensional terms have been used:

$$X = \frac{x}{L},$$

$$W = \frac{w}{L},$$

$$\alpha = \frac{e_0 a}{L} = \text{scaling effect parameter},$$

$$\bar{N}^0 = \frac{\bar{N}L^2}{EI_0} = \text{buckling load parameter},$$

$$\Omega = \frac{EI_0}{k_s GAL^2},$$

$$\bar{\Omega} = \frac{G\tilde{A}L^2}{EI_0},$$

$$\hat{\Omega} = \frac{EI_0}{G\tilde{A}L^2}.$$

Below we have shown the non-dimensionalized forms of the governing differential equations for EBT, TBT, RBT, and LBT in Eqs. (5.1)–(5.4), respectively.

$$e^{-\eta X}\frac{d^4W}{dX^4} = \bar{N}^0\left(\alpha^2\frac{d^4W}{dX^4} - \frac{d^2W}{dX^2}\right), \tag{5.1}$$

$$e^{-\eta X}\frac{d^4W}{dX^4} = \bar{N}^0\left(\Omega e^{-\eta X}\frac{d^4W}{dX^4} - \Omega e^{-\eta X}\alpha^2\frac{d^6W}{dX^6} - \frac{d^2W}{dX^2} + \alpha^2\frac{d^4W}{dX^4}\right), \tag{5.2}$$

$$\frac{105}{84}\bar{\Omega}\frac{d^4W}{dx^4} - \frac{1}{105}e^{-\eta X}\frac{d^6W}{dx^6}$$
$$= \bar{N}^0\left(-\frac{105}{84}\bar{\Omega}e^{\eta X}\frac{d^2W}{dX^2} + \frac{105}{84}\bar{\Omega}e^{\eta X}\alpha^2\frac{d^4W}{dX^4}\right.$$
$$\left. + \frac{68}{84}\frac{d^4W}{dX^4} - \frac{68}{84}\alpha^2\frac{d^6W}{dX^6}\right), \tag{5.3}$$

$$e^{-\eta X}\frac{d^4W}{dX^4} = \bar{N}^0\left(\frac{4}{5}\hat{\Omega}e^{-\eta X}\frac{d^4W}{dX^4} - \frac{4}{5}\hat{\Omega}e^{-\eta X}\alpha^2\frac{d^6W}{dX^6} - \frac{d^2W}{dX^2} + \alpha^2\frac{d^4W}{dX^4}\right). \tag{5.4}$$

By the application of DQM, one may obtain generalized eigenvalue problem as

$$[K]\{W\} = \bar{N}^0 [B_c]\{W\}, \tag{5.5}$$

where K is the stiffness matrix and B_c is the buckling matrix.

5.1 Numerical Results and Discussions

In this section, numerical results have been computed by solving Eq. (5.5) using MATLAB program developed by the authors. DQM has been employed and boundary conditions are implemented in the coefficient matrices. Here unless mentioned, buckling load would denote the critical buckling load parameter (first eigenvalue). It may be noted that the following parameters are taken for the computation (Sahmani and Ansari 2011; Reddy 2007): $E = 70$ GPa, $\upsilon = 0.23$, $h = 1$, $k_s = 5/6$.

5.1.1 *Convergence*

To find the minimum number of grid points for obtaining desired results, a convergence study has been carried out for EBT and TBT nanobeams. To show how the solution is being affected by grid points, variation of critical buckling load parameter with number grid points (n) has been shown in Fig. 5.1. In this figure, we have taken $e_0a = 1$ nm, non-uniform parameter $(\eta) = 0.2$ and $L = 10$ nm. The convergence has been shown for simply supported edge condition only. From this figure, one may observe that convergence is achieved as we increase the number of grid points. It may be noted that 11 grid points are sufficient to compute the results.

5.1.2 *Validation*

To validate the present results, we compare our results with that of available in the literature. For the validation purpose, we consider an uniform ($\eta = 0$) nanobeam. To compare our results of EBT and TBT nanobeams with Wang *et al.* (2006), we have considered a beam of diameter $d = 1$ nm, Young's modulus $E = 1$ TPa, and Poisson's ratio $\upsilon = 0.19$. The comparison has been shown in Table 5.1 for three types of boundary conditions such as S–S, C–S, and C–C. In this table, the critical buckling load parameter (in nN) for EBT and TBT nanobeams with $L/d = 10$ have been presented for

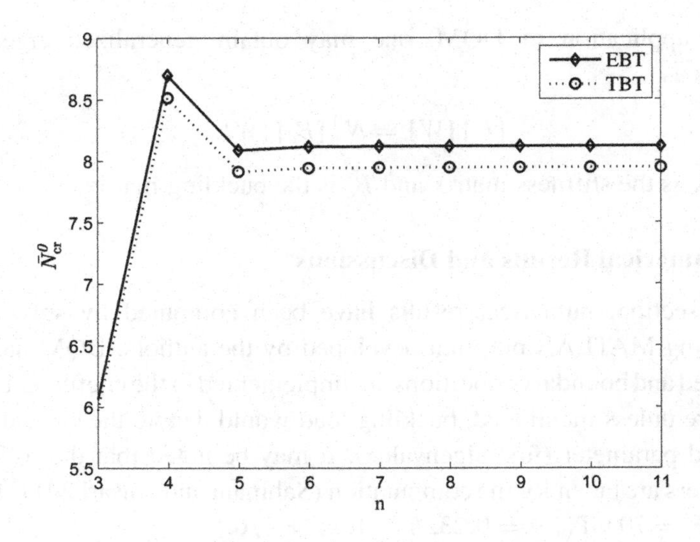

Fig. 5.1 Variation of $\bar{N}^0_{cr}$ with grid points

Table 5.1 Comparison of critical buckling load parameter $\bar{N}^0_{cr}$ (nN) for EBT and TBT nanobeams

	L/d	e_0a	EBT		TBT	
			Present	Ref.*	Present	Ref.*
S–S	10	0	4.8447	4.8447	4.7835	4.7670
		0.5	4.7280	4.7281	4.6683	4.654
		1	4.4095	4.4095	4.3534	4.3450
C–S	10	0	9.91109	9.9155	9.5580	9.5605
		0.5	9.4348	9.4349	9.1934	9.1179
		1	8.2461	8.2461	8.0356	8.0055
C–C	10	0	19.3789	19.379	18.4342	18.192
		0.5	17.6381	17.6381	16.7783	16.649
		1	13.8939	13.8939	13.2165	13.273

*Wang *et al.* (2006).

various values of scale coefficients (0, 0.5, 1 nm). Similarly, buckling load of RBT and LBT nanobeams have been compared, respectively, with Emam (2013) and Sahmani and Ansari (2011) in Table 5.2. It may be noted that comparison for RBT nanobeams has been made with aspect ratio (*L/h*) as

Table 5.2 Comparison of critical buckling load parameter ($\bar{N}_{cr}^{0}$) for RBT and LBT nanobeams

		S–S		C–C	
L/h	μ	Present	Ref.*	Present	Ref.*
RBT					
10	0	9.6228	9.6228	35.8075	35.8075
	1	8.7583	8.7583	25.6724	25.6724
	2	8.0364	8.0364	20.0090	20.0090
	3	7.4245	7.4245	16.3927	16.3927
LBT					
50	0	9.8595	9.8616	39.4170	39.4457
	0.5	9.8401	9.8422	39.3589	39.3899
	1	9.8207	9.8228	38.2958	39.3118
	1.5	9.8014	9.8036	38.2072	39.2351

*Emam (2013).
**Sahmani and Ansari (2011).

10 and nonlocal parameter (μ) as 0, 1, 2, 3 nm^2, while comparison for LBT nanobeams has been done with $L/h = 50$ and $\mu = 0, 0.5, 1, 1.5$ nm^2. In this table, we have considered S–S and C–C edge conditions. It is seen that critical buckling load parameter ($\bar{N}_{cr}^{0}$) decreases with increase in nonlocal parameter. From these tables, one may observe close agreement of results with that of available in the literatures.

5.1.3 *Effect of small scale*

In this section, the significance of scale coefficient has been highlighted. First, we define buckling load ratio as $\bar{N}^0$ calculated using nonlocal theory/ $\bar{N}^0$ calculated using local theory. This buckling load ratio serves as an index to estimate quantitatively the small scale effect on the buckling solution. To state the importance of scale coefficient, variation of buckling load ratio (associated with first mode) with scale coefficient ($e_0 l_{int}$) has been shown in Figs. 5.2–5.5. It may be noted that Figs. 5.2 and 5.3 present graphical results for RBT nanobeams with guided and simply supported-guided, respectively.

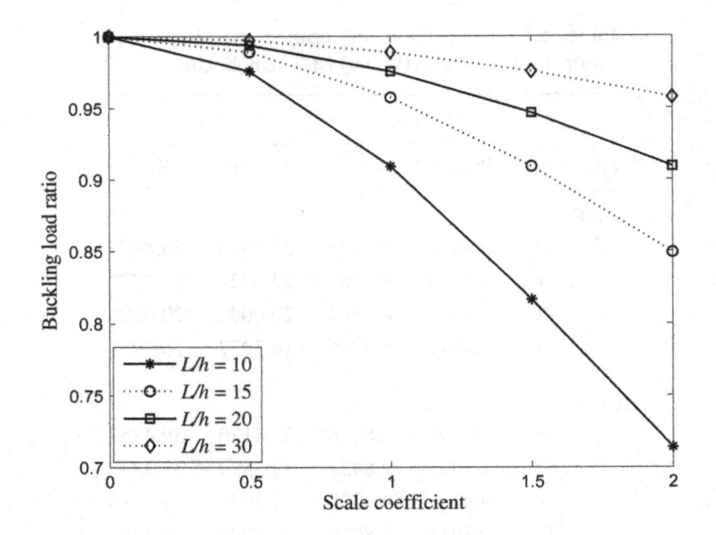

Fig. 5.2 Variation of buckling load ratio with $e_0 l_{int}$

Similarly, Figs. 5.4 and 5.5 illustrate results for LBT nanobeams with guided and simply supported-guided, respectively. In these figures, we have taken non-uniform parameter (η) as 0.4. Results have been shown for different values of aspect ratio (L/h). It is noticed from the figures that buckling load ratios are less than unity. This implies that application of local beam model for the buckling analysis of carbon nanotubes (CNTs) would lead to overprediction of the buckling load if the small length-scale effect between the individual carbon atoms is neglected. As the scale coefficient ($e_0 l_{int}$) increases, buckling loads obtained by nonlocal beam model become smaller than those of its local counterpart. In other words, buckling load parameter obtained by local beam theory is more than that obtained by nonlocal beam theory. So, the presence of nonlocal parameter in the constitutive equation is significant in the field of nanomechanics. It is also observed that the small scale effect is affected by L/h. This observation is explained as: when L/h increases, buckling load ratio comes closer to 1. This implies that buckling load parameter obtained by nonlocal beam model comes closer to that furnished by local beam model. Hence, small scale effect is negligible for a very slender CNT while it is significant for short CNTs. This implies that if we compare magnitude of small scale effect with length of the slender

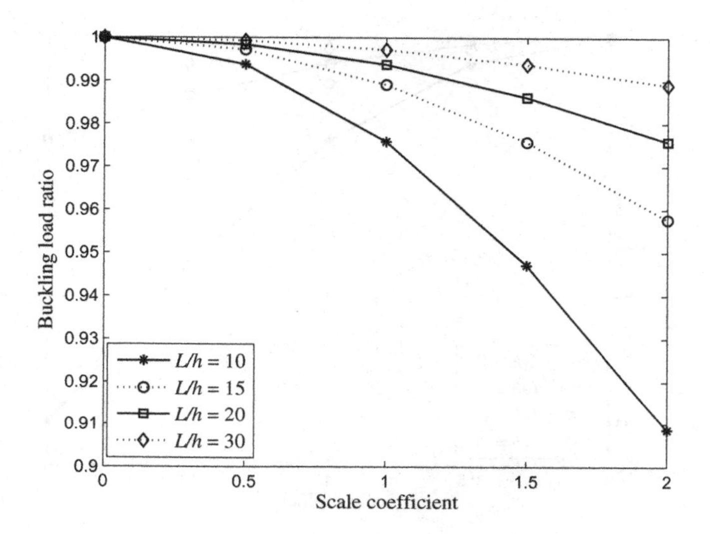

Fig. 5.3 Variation of buckling load ratio with $e_0 l_{\mathrm{int}}$

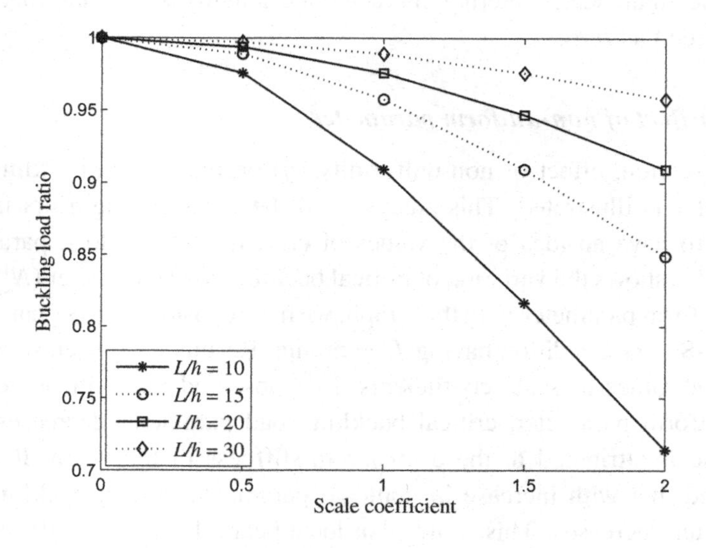

Fig. 5.4 Variation of buckling load ratio with $e_0 l_{\mathrm{int}}$

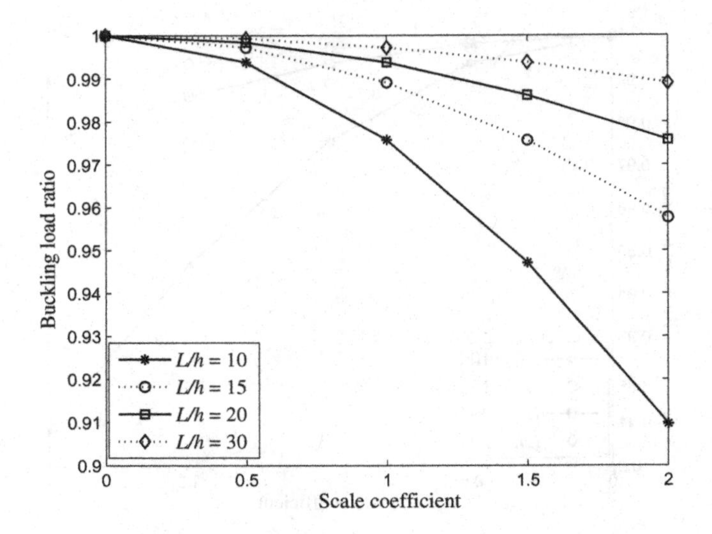

Fig. 5.5 Variation of buckling load ratio with $e_0 l_{\text{int}}$

tube, the small scale (internal characteristic length) is so small that it can be regarded as zero.

5.1.4 *Effect of non-uniform parameter*

In this section, effect of non-uniformity (η) on the critical buckling load parameter is illustrated. This analysis will help design engineers in their design to have an idea of the values of critical buckling load parameter. Figure 5.6 shows the variation of critical buckling load parameter ($\bar{N}^0_{\text{cr}}$) with non-uniform parameter η. In this graph, we have considered EBT nanobeam with C–S edge condition having $L = 50$ nm. Results have been shown for different values of scale coefficients. It is observed that with increase in non-uniform parameter, critical buckling load parameter decreases. This decrease is attributed to the decrease in stiffness of the beam. It is also observed that with increase in nonlocal parameter, critical buckling load parameter decreases. This shows that local beam theory ($\mu = 0$) overpredicts buckling load parameter. Hence, for better predictions of buckling load parameter, one should consider nonlocal theory. Next, to investigate the influence of non-uniform parameter on the higher buckling modes, variation of buckling load parameter ($\bar{N}^0$) with non-uniform parameter (η) has

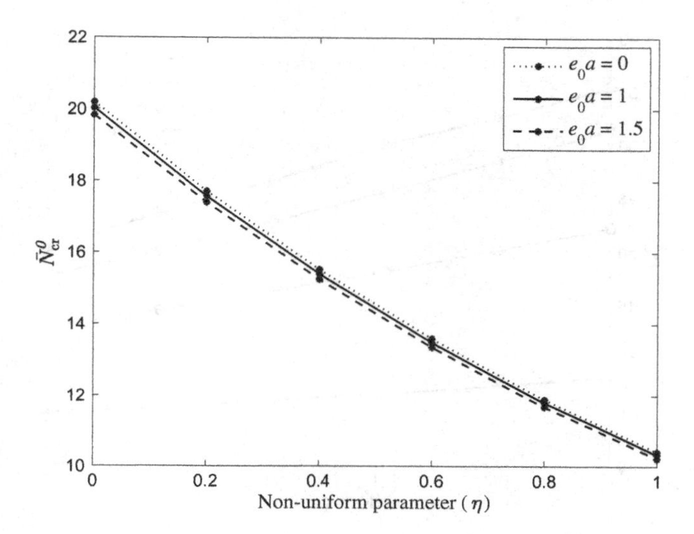

Fig. 5.6 Variation of $\bar{N}^0_{\mathrm{cr}}$ with non-uniform parameter

been presented in Fig. 5.7. Here we have considered LBT nanobeams with S–S edge condition having $\mu = 2\ \mathrm{nm}^2$ and $L = 15$ nm. It is seen from the figure that buckling load parameters decrease with increase in non-uniform parameter and this decrease is more significant in the case of higher modes.

5.1.5 *Effect of aspect ratio*

One of the another important factors that design engineers should keep in mind is that the effect of aspect ratio. To investigate the effect of aspect ratio (*L/h*) on the critical buckling load parameter, variation of critical buckling load parameter with *L/h* has been shown in Figs. 5.8–5.11. Results have been shown for different values of scale coefficients (0.5, 1, 1.5 nm). In these graphs, we have considered EBT and TBT nanobeams with edge conditions such as C–S and C–C. Numerical values have been obtained taking $\eta = 0.5$. It is observed that buckling load increases with increase in *L/h*. It is also noticed that buckling load decreases with increase in scale coefficient. Hence one should incorporate nonlocal theory in the buckling analysis of nanobeams. One may also notice that for a particular value of *L/h*, results obtained by TBT nanobeams are less as compared to EBT nanobeams. This is due to the absence of transverse shear stress and strain

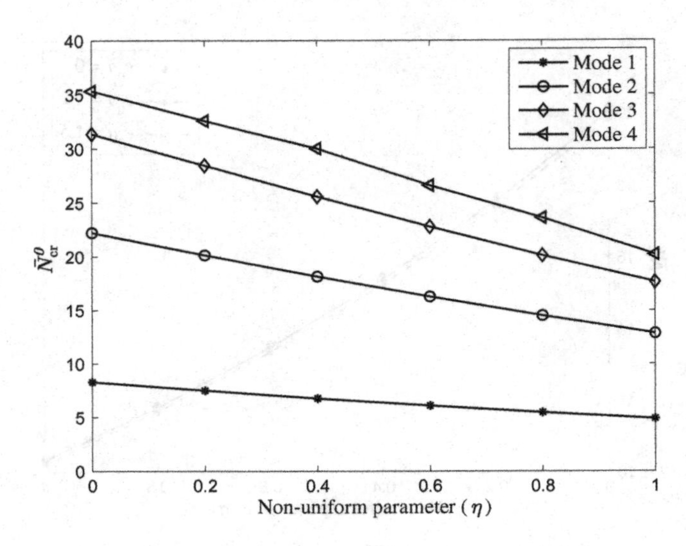

Fig. 5.7 Variation of $\bar{N}^0$ with non-uniform parameter

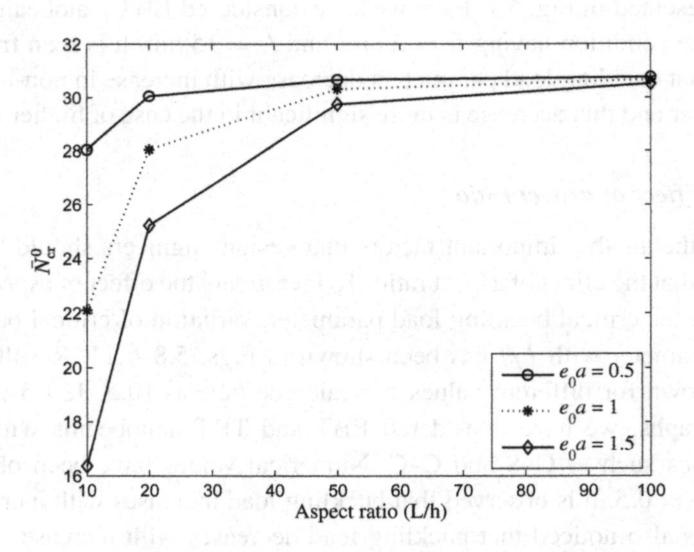

Fig. 5.8 Variation of $\bar{N}_{cr}^{0}$ with *L/h* (EBT C–C)

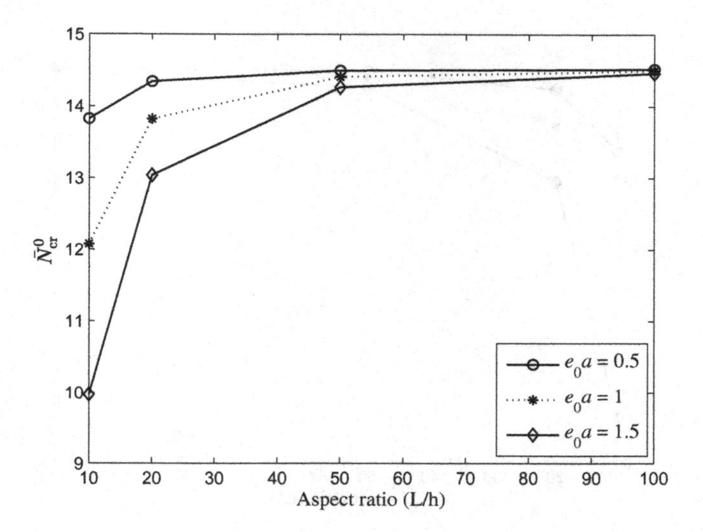

Fig. 5.9 Variation of $\bar{N}_{cr}^{0}$ with *L/h* (EBT C–S)

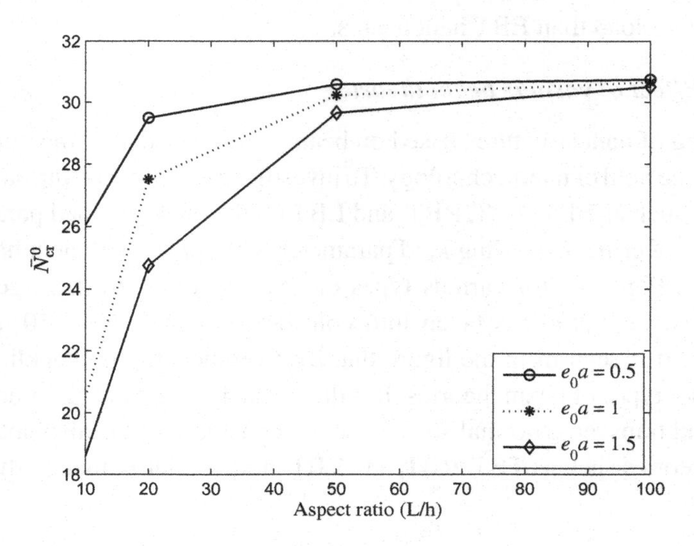

Fig. 5.10 Variation of $\bar{N}_{cr}^{0}$ with *L/h* (TBT C–C)

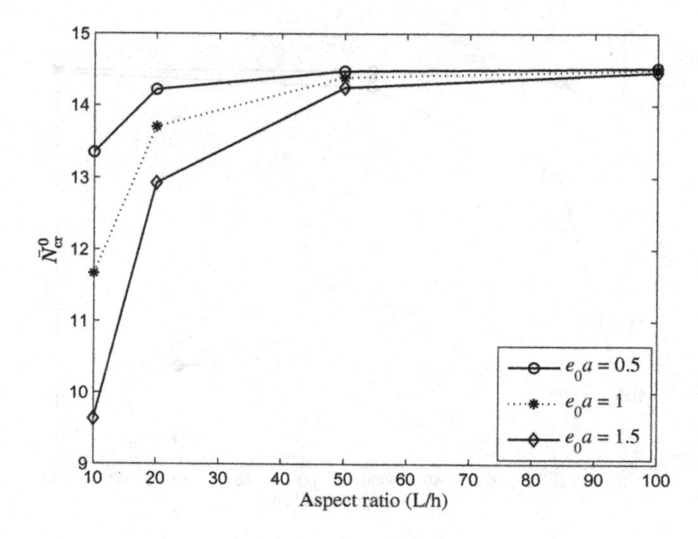

Fig. 5.11 Variation of $\bar{N}^0_{cr}$ with L/h (TBT C–S)

in EBT nanobeams. One may say TBT nanobeams predict better prediction of buckling load than EBT nanobeams.

5.1.6 *Effect of various beam theories*

Modeling of nanostructures based on beam theories is one of the important areas in the field of nanotechnology. To investigate the effect of various beam theories such as EBT, TBT, RBT, and LBT on the buckling load parameter, variation of critical buckling load parameter with scale coefficient has been shown in Fig. 5.12 for various types of beam theories. In this figure, C–S boundary condition is taken into consideration with $L = 10$ nm and $\eta = 0.5$. It is seen from the figure that EBT predicts higher buckling load than other types of beam theories. It is due to the fact that in EBT, transverse shear and transverse normal strains are not considered. It is also noted that beam theories such as TBT, RBT, and LBT predict approximately the same results.

5.1.7 *Effect of boundary condition*

For designing engineering structures, one must have proper knowledge about boundary conditions. It will help engineers to have an idea without

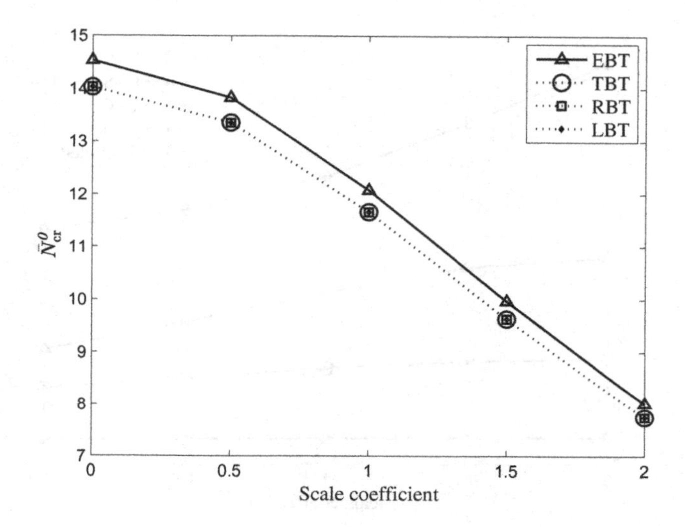

Fig. 5.12 Variation of $\bar{N}^0_{cr}$ with scale coefficient

carrying out detailed investigation. Therefore, analysis of boundary conditions is quite important. In this section, we have considered the effect of boundary condition on the critical buckling load parameter. Figure 5.13 depicts variation of critical buckling load parameter of TBT nanobeam with scale coefficient for different boundary conditions. This graph is plotted with $L = 10$ nm and $\eta = 0.2$. It is observed from the figure that C–C nanobeams are having highest critical buckling load parameter and simply supported-guided nanobeams are having lowest critical buckling load parameter.

Buckling of embedded nanobeams

Here, we have investigated buckling of embedded nanobeams in thermal environments based on EBT, TBT, and RBT. The nanobeam is embedded in elastic foundations such as Winkler and Pasternak. Rayleigh–Ritz has been applied in EBT and TBT with shape functions as boundary characteristic orthogonal polynomials and Chebyshev polynomials, respectively. DQM has been employed in buckling of embedded nanobeams based on RBT.

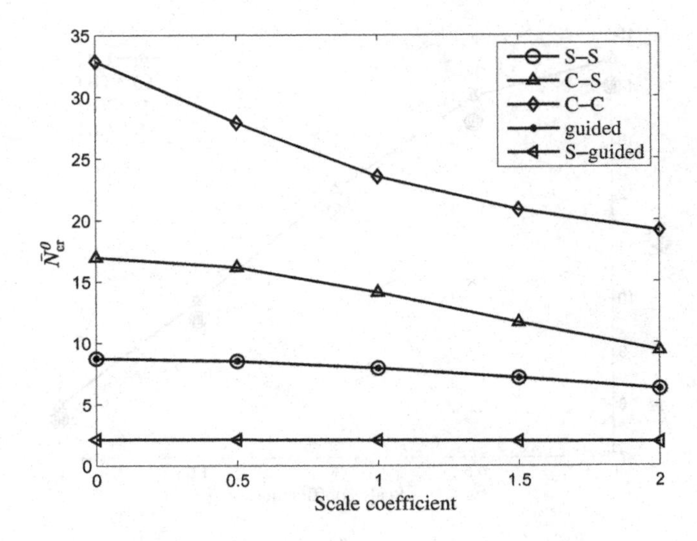

Fig. 5.13 Variation of $\bar{N}^0_{cr}$ with scale coefficient

For simplicity and convenience in mathematical formulation, following non-dimensional parameters have been introduced here:

$$X = \frac{x}{L}, \quad W = \frac{w}{L}, \quad \alpha = \frac{e_0 a}{L}, \quad \hat{N} = \frac{\bar{N} L^2}{EI},$$

$$\tau = \frac{I}{AL^2}, \quad \Omega = \frac{EI}{k_s G A L^2}, \quad \bar{\Omega} = \frac{G\bar{A}L^2}{EI}, \quad K_g = \frac{k_g L^2}{EI},$$

$$K_w = \frac{k_w L^4}{EI}, \quad \hat{N}_\theta = \frac{N_\theta L^2}{EI}, \quad \hat{N}_m = \frac{N_m L^2}{EI}.$$

5.1.7.1 *Euler–Bernoulli beam theory (EBT)*

The strain energy may be written as Eq. (3.1).

The potential energy due to the axial force may be expressed as (Amirian *et al.* 2013)

$$V_a = \frac{1}{2} \int_0^L \left\{ \bar{N} \left(\frac{dw}{dx} \right)^2 + f_e w \right\} dx, \tag{5.6}$$

where $\bar{N}$ is the axial force and is expressed as $\bar{N} = N_m + N_\theta$. Here N_m is the axial force due to the mechanical loading prior to buckling and N_θ is the axial force due to the influence of temperature change which is defined by $N_\theta = -\frac{EA}{1-2\nu}\alpha_x\theta$.

In this expression, E is Young's modulus, A the cross-sectional area, ν Poisson's ratio, α_x the coefficient of thermal expansion in the direction of x axis, and θ is the change in temperature. Also f_e is the density of reaction force of elastic foundation which is expressed as $f_e = k_w w - k_g \frac{d^2w}{dx^2}$, where k_w denotes Winkler modulus and k_g denotes shear modulus of the elastic medium.

Using Hamilton's principle and setting coefficient of δw to zero, we obtain the following governing equation of motion:

$$\frac{d^2M}{dx^2} + \bar{N}\frac{d^2w}{dx^2} - k_w w + k_g \frac{d^2w}{dx^2} = 0. \tag{5.7}$$

Using Eqs. (1.7) and (5.7), M in nonlocal form may be written as

$$M = -EI\frac{d^2w}{dx^2} + \mu\left[-\bar{N}\frac{d^2w}{dx^2} + k_w w - k_g\frac{d^2w}{dx^2}\right].$$

Let us substitute $N_m = -P_b$.

Equating energies of the system, we may obtain Rayleigh quotient ($\bar{N}^0$) from the following eigen equation in non-dimensional form:

$$\begin{aligned}\bar{N}^0&\left[\left(\frac{dW}{dX}\right)^2 + \alpha^2\left(\frac{d^2W}{dX^2}\right)^2\right]\\ &= \left(\frac{d^2W}{dX^2}\right)^2 - K_w\alpha^2 W\frac{d^2W}{dX^2} + K_g\alpha^2\left(\frac{d^2W}{dX^2}\right)^2\\ &\quad + K_w W^2 - K_g W\frac{d^2W}{dX^2} + \hat{N}_\theta\left(\frac{dW}{dX}\right)^2 + \hat{N}_\theta\alpha^2\left(\frac{d^2W}{dX^2}\right)^2,\end{aligned} \tag{5.8}$$

where $\bar{N}^0 = P_b L^2/EI$ is the non-dimensional buckling load parameter.

Here, we have used orthonormal polynomials ($\hat{\varphi}_k$) in Eq. (3.5). Substituting Eq. (3.5) in Eq. (5.8) and minimizing $\bar{N}^0$ with respect to constant coefficients, the following eigenvalue value problem may be obtained:

$$[K]\{Z\} = \bar{N}^0 [B_c]\{Z\}, \tag{5.9}$$

where Z is a column vector of constants, stiffness matrix K and buckling matrix B_c are given as below:

$$K(i,j) = \int_0^1 ((2 + 2K_g\alpha^2 + 2\hat{N}_\theta\alpha^2)\phi_i''\phi_j'' - K_w\alpha^2\phi_i''\phi_j - K_w\alpha^2\phi_i\phi_j'' + 2K_w\phi_i\phi_j - K_g\phi_i''\phi_j - K_g\phi_i\phi_j'' + 2\hat{N}_\theta\alpha^2\phi_i'\phi_j')dX,$$

$$B_c(i,j) = \int_0^1 (2\phi_i'\phi_j' + 2\alpha^2\phi_i''\phi_j'')dX.$$

5.1.7.2 *Timoshenko beam theory (TBT)*

The strain energy may be given as Eq. (3.9).

The potential energy due to the axial force may be expressed as Eq. (5.6).

Using Hamilton's principle, governing equations are obtained as

$$\frac{dM}{dx} - Q = 0, \tag{5.10}$$

$$\frac{dQ}{dx} + \bar{N}\frac{d^2w}{dx^2} - f_e = 0. \tag{5.11}$$

Using Eqs. (5.10) and (5.11) and Eqs. (1.13) and (1.14), one may obtain bending moment M and shear force Q in nonlocal form as follows:

$$M = EI\frac{d\phi}{dx} + \mu\left[-\bar{N}\frac{d^2w}{dx^2} + f_e\right], \tag{5.12}$$

$$Q = k_s GA\left(\phi + \frac{dw}{dx}\right) + \mu\left[-\bar{N}\frac{d^3w}{dx^3} + k_w\frac{dw}{dx} - k_g\frac{d^3w}{dx^3}\right]. \tag{5.13}$$

Equating energies of the system, one may obtain following expressions for TBT nanobeams in non-dimensional form:

$$\bar{N}^0\left[\left(\frac{dW}{dX}\right)^2 - \alpha^2\frac{d\phi}{dX}\frac{d^2W}{dX^2} - \alpha^2\frac{d^3W}{dX^3}\left(\phi + \frac{dW}{dX}\right)\right]$$

$$= \left(\frac{d\phi}{dX}\right)^2 - \hat{N}_\theta \alpha^2 \frac{d\phi}{dX}\frac{d^2W}{dX^2} - K_w \alpha^2 W \frac{d\phi}{dX}$$
$$- K_g \alpha^2 \frac{d^2W}{dX^2}\frac{d\phi}{dX} + \frac{1}{\Omega}\left(\phi + \frac{dW}{dX}\right)^2$$
$$- \hat{N}_\theta \alpha^2 \frac{d^3W}{dX^3}\left(\phi + \frac{dW}{dX}\right) + K_w \alpha^2 \frac{dW}{dX}\left(\phi + \frac{dW}{dX}\right)$$
$$- K_g \alpha^2 \frac{d^3W}{dX^3}\left(\phi + \frac{dW}{dX}\right) + \hat{N}_\theta \left(\frac{dW}{dX}\right)^2$$
$$+ K_w W^2 - K_g W \frac{d^2W}{dX^2}. \tag{5.14}$$

As the present technique is applicable only in the interval -1 to 1, we introduce another independent variable ξ as $\xi = 2X - 1$ which transforms the range $0 \leq X \leq 1$ into the applicability range $-1 \leq \xi \leq 1$.

Substituting $N_m = -P_b$ and equating energies of the system, one may obtain Rayleigh quotient ($\bar{N}^0$) from following equation:

$$\bar{N}^0 \left[4\left(\frac{dW}{d\xi}\right)^2 - 8\alpha^2 \frac{d\phi}{d\xi}\frac{d^2W}{d\xi^2} - 8\alpha^2 \frac{d^3W}{d\xi^3}\left(\phi + 2\frac{dW}{d\xi}\right)\right]$$
$$= 4\left(\frac{d\phi}{d\xi}\right)^2 - 8\hat{N}_\theta \alpha^2 \frac{d\phi}{d\xi}\frac{d^2W}{d\xi^2} + 2K_w \alpha^2 W \frac{d\phi}{d\xi} - 8K_g \alpha^2 \frac{d^2W}{d\xi^2}\frac{d\phi}{d\xi}$$
$$+ \frac{1}{\Omega}\left(\phi + 2\frac{dW}{d\xi}\right)^2 - 8\hat{N}_\theta \alpha^2 \frac{d^3W}{d\xi^3}\left(\phi + 2\frac{dW}{d\xi}\right)$$
$$+ 2K_w \alpha^2 \frac{dW}{d\xi}\left(\phi + 2\frac{dW}{d\xi}\right) - 8K_g \alpha^2 \frac{d^3W}{d\xi^3}\left(\phi + 2\frac{dW}{d\xi}\right)$$
$$+ 4\hat{N}_\theta \left(\frac{dW}{d\xi}\right)^2 + K_w W^2 - 4K_g W \frac{d^2W}{d\xi^2}, \tag{5.15}$$

where $\bar{N}^0 = P_b L^2/EI$ is the buckling load parameter.

Substituting Eqs. (3.14) and (3.15) in Eq. (5.15) and minimizing $\bar{N}^0$ with respect to the constant coefficients, the following eigenvalue value problem is obtained:

$$[K]\{Z\} = \bar{N}^0 [B_c]\{Z\}, \tag{5.16}$$

where Z is a column vector of constants. Here K and B_c are the stiffness and buckling matrices for TBT nanobeams which are given by:

$$K = \begin{bmatrix} k_1 & k_2 \\ k_3 & k_4 \end{bmatrix},$$

where k_1, k_2, k_3, and k_4 are submatrices and are given as

$k_1(i,j) = \int_{-1}^{1} (8(\frac{1}{\Omega} + K_w\alpha^2 + \hat{N}_\theta)\varphi_i'\varphi_j' - 16(\hat{N}_\theta\alpha^2 + K_g\alpha^2)\varphi_i'''\varphi_j'$
$-16(\hat{N}_\theta\alpha^2 + K_g\alpha^2)\varphi_i'\varphi_j''' + 2K_w\varphi_i\varphi_j - 4K_g\varphi_i''\varphi_j - 4K_g\varphi_i\varphi_j'')d\xi,$

$k_2(i,j) = \int_{-1}^{1} (-8\hat{N}_\theta\alpha^2\varphi_i''\psi_j' + 2K_w\alpha^2\varphi_i\psi_j' - 8K_g\alpha^2\varphi_i''\psi_j' + 4\frac{1}{\Omega}\varphi_i'\psi_j$
$-(8\hat{N}_\theta\alpha^2 + 8K_g\alpha^2)\varphi_i'''\psi_j + 2K_w\alpha^2\varphi_i'\psi_j)d\xi,$

$k_3(i,j) = \int_{-1}^{1} (-8\hat{N}_\theta\alpha^2\psi_i'\varphi_j'' + 2K_w\alpha^2\psi_i'\varphi_j - 8K_g\alpha^2\psi_i'\varphi_j'' + 4\frac{1}{\Omega}\psi_i\varphi_j'$
$-8\hat{N}_\theta\alpha^2\psi_i\varphi_j''' + 2K_w\alpha^2\psi_i\varphi_j' - 8K_g\alpha^2\psi_i\varphi_j''')d\xi,$

$k_4(i,j) = \int_{-1}^{1}(8\psi_i'\psi_j' + 2\frac{1}{\Omega}\psi_i\psi_j)d\xi,$

$$B_c = \begin{bmatrix} B_1 & B_2 \\ B_3 & B_4 \end{bmatrix},$$

where B_1, B_2, B_3, and B_4 are submatrices and are given as

$B_1(i,j) = \int_{-1}^{1}(8\varphi_i'\varphi_j' - 16\alpha^2\varphi_i'\varphi_j''' - 16\alpha^2\varphi_i'''\varphi_j')d\xi,$
$B_2(i,j) = \int_{-1}^{1}(-8\alpha^2\varphi_i''\psi_j' - \alpha^2\varphi_i'''\psi_j)d\xi,$
$B_3(i,j) = \int_{-1}^{1}(-8\alpha^2\psi_i'\varphi_j'' - 8\alpha^2\psi_i\varphi_j''')d\xi,$
$B_4(i,j) = 0.$

5.1.7.3 *Reddy beam theory (RBT)*

Governing equations of embedded nanobeams based on RBT may be written as (Reddy 2007)

$$G\tilde{A}\left(\frac{d\phi}{dx} + \frac{d^2w}{dx^2}\right) - \bar{N}\frac{d^2w}{dx^2} - k_w w + k_g\frac{d^2w}{dx^2}$$
$$+\mu\left[\bar{N}\frac{d^4w}{dx^4} + k_w\frac{d^2w}{dx^2} - k_g\frac{d^4w}{dx^4}\right]$$
$$+c_1EJ\frac{d^3\phi}{dx^3} - c_1^2EK\left(\frac{d^3\phi}{dx^3} + \frac{d^4w}{dx^4}\right) = 0, \tag{5.17}$$

$$E\hat{I}\frac{d^2\phi}{dx^2} - c_1E\hat{J}\left(\frac{d^2\phi}{dx^2} + \frac{d^3w}{dx^3}\right) - G\tilde{A}\left(\phi + \frac{dw}{dx}\right) = 0. \tag{5.18}$$

Eliminating ϕ from Eqs. (5.17) and (5.18), governing equations may be obtained in terms of displacement as

$$-\left(\frac{68}{84}\bar{N}+\frac{105}{84EI}G\tilde{A}\mu\bar{N}\right)\frac{d^4w}{dx^4}+\frac{105}{84EI}G\tilde{A}\bar{N}\frac{d^2w}{dx^2}+\frac{68}{84}\mu\bar{N}\frac{d^6w}{dx^6}$$
$$=\left(\frac{68}{84}k_w+\frac{105}{84EI}G\tilde{A}k_g+\frac{105}{84EI}G\tilde{A}\mu k_w\right)\frac{d^2w}{dx^2}$$
$$-\left(\frac{68}{84}k_g+\frac{68}{84}\mu k_w+\frac{105}{84EI}G\tilde{A}\mu k_g+\frac{21}{84}G\tilde{A}\right)\frac{d^4w}{dx^4}$$
$$+\left(\frac{68}{84}\mu k_g+\frac{1}{105}EI\right)\frac{d^6w}{dx^6}-\frac{105}{84EI}G\tilde{A}k_w w. \tag{5.19}$$

Substituting $N_m = -P_b$, governing equation in non-dimensional form is obtained as

$$\bar{N}^0\left[\left(\frac{68}{84}-\frac{105}{84}\bar{\Omega}\alpha^2\right)\frac{d^4W}{dX^4}-\frac{105}{84}\bar{\Omega}\frac{d^2W}{dX^2}-\frac{68}{84}\alpha^2\frac{d^6W}{dX^6}\right]$$
$$=\left(\frac{68}{84}K_w+\frac{105}{84}\bar{\Omega}K_g+\frac{105}{84}\bar{\Omega}\alpha^2K_w+\frac{105}{84}\bar{\Omega}\hat{N}_\theta\right)\frac{d^2W}{dX^2}$$
$$+\left(-\frac{68}{84}K_g-\frac{68}{84}\alpha^2K_w-\frac{105}{84}\bar{\Omega}\alpha^2K_g-\frac{21}{84}\bar{\Omega}\right.$$
$$\left.-\hat{N}_\theta\frac{68}{84}+\frac{105}{84}\bar{\Omega}\alpha^2\hat{N}_\theta\right)\frac{d^4W}{dX^4}$$
$$+\left(\frac{68}{84}\alpha^2K_g+\frac{1}{105}+\frac{68}{84}\alpha^2\hat{N}_\theta\right)\frac{d^6W}{dX^6}-\frac{105}{84}\bar{\Omega}K_w, \tag{5.20}$$

where $\bar{N}^0 = P_bL^2/EI$ is the buckling load parameter.

Application of DQM in Eq. (5.20), one may obtain generalized eigenvalue problem as

$$[K]\{W\}=\bar{N}^0\,[B_c]\{W\}, \tag{5.21}$$

where K is the stiffness matrix and B_c is the buckling matrix.

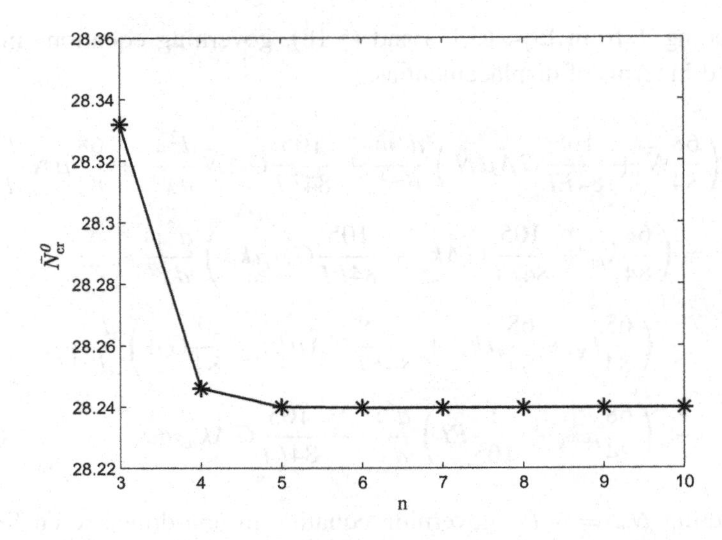

Fig. 5.14 Convergence of critical buckling load parameter (EBT)

5.2 Numerical Results and Discussions

Buckling of single-walled carbon nanotubes (SWCNTs) embedded in elastic medium including thermal effect has been investigated. The elastic medium is modeled as Winkler-type and Pasternak-type foundations. The effective properties of SWCNTs are taken as follows (Benzair *et al.* 2008; Murmu and Pradhan 2009b): Young's modulus (E) $=$ 1000 GPa, Poisson's ratio (ν) $= 0.19$, shear correction factor (k_s) $= 0.877$, $\alpha_x = -1.6 \times 10^{-6}$ K^{-1} for room or low temperature and $\alpha_x = 1.1 \times 10^{-6}$ K^{-1} for high temperature. A computer code is developed by the authors in MATLAB based on Eqs. (5.9), (5.16), and (5.21).

5.2.1 *Convergence*

First of all, convergence test has been performed to find minimum number of terms required for computation. As such, Figs. 5.14 and 5.15 illustrate convergence of critical buckling load parameter ($\bar{N}_{cr}^{0}$), respectively, for EBT and RBT. In Fig. 5.14, we have considered C–S edge condition with $L/h = 20$, $e_0a = 1$ nm, $K_w = 60$, $K_g = 4$, $\theta = 10$ K, and in Fig. 5.15, we have taken $K_w = 50$, $K_g = 2$, $\theta = 20$ K, $e_0a = 1.5$ nm, $L/h = 20$ with C–S support. Similarly, Table 5.3 shows convergence of critical buckling

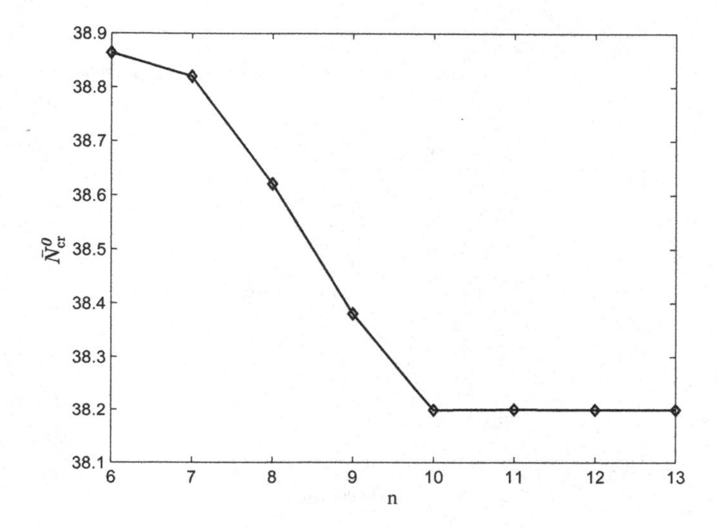

Fig. 5.15 Convergence of critical buckling load parameter (RBT)

Table 5.3 Convergence of first three buckling load parameters (TBT)

n	First	Second	Third
4	22.2348	39.7800	65.3622
5	22.1742	38.0072	51.6357
6	22.1711	37.6160	49.2176
7	22.1600	37.6100	47.9400
8	22.1616	37.4205	43.5535
9	22.1500	35.8300	37.7600
9	22.1500	35.8300	37.7600

load parameter of nanobeams based on TBT. In this table, we have taken $K_w = 50$, $K_g = 2$, $\theta = 10$ K, $e_0a = 1$ nm, $L/h = 10$, and C–S edge condition. One may note that convergence test has been performed in low-temperature environment. Above convergence patterns show that 10 grid points are sufficient to obtain results in the present analysis.

5.2.2 *Validation*

To validate the present results, a comparison study has been carried out with the results of Wang *et al.* (2006). For this comparison, we have taken

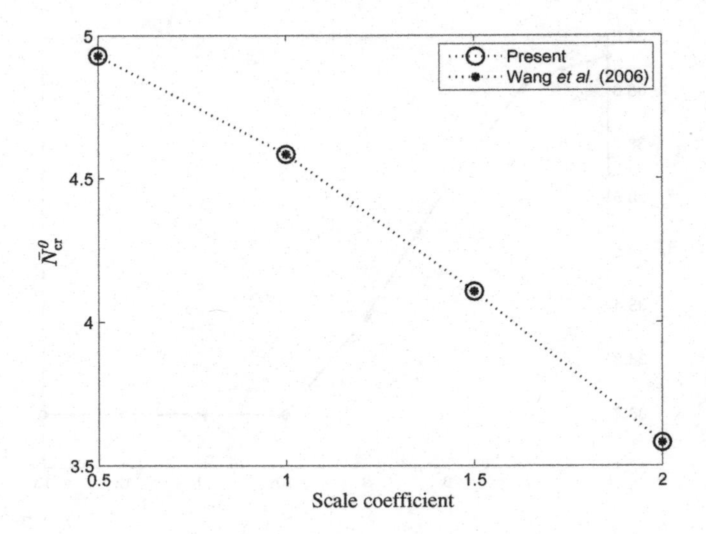

Fig. 5.16 Comparison of critical buckling load parameter (EBT)

$K_w = 0$, $K_g = 0$, and $\theta = 0$ K. As such, Figs. 5.16 and 5.17 show graphical comparisons of EBT and TBT nanobeams, respectively. In Fig. 5.16, we have considered C–S support with $L/d = 14$ and in Fig. 5.17, we have considered S–S support with $L/d = 10$. Similarly, tabular comparison study has been tabulated in Table 5.4 for RBT nanobeams with that of Emam (2013) for $L/h = 10$. For this comparison, we have taken same parameters as that of Emam (2013). One may find a close agreement of the results. This shows the suitability and reliability of the present method for the buckling analyses of SWCNTs.

5.2.3 *Effect of Winkler modulus parameter*

In this section, we have investigated the influence of surrounding medium on the buckling analysis of SWCNTs. The elastic medium is modeled as Winkler-type and Pasternak-type foundations. Figures 5.18–5.20 illustrate effect of Winkler modulus parameter on the buckling solutions based on EBT, TBT, and RBT, respectively. We have shown these graphical results in low-temperature environment with $K_g = 0$. Numerical values taken for this computation are $\theta = 30$ K, $L/h = 10$ in Fig. 5.18 with C–F support, whereas in Fig. 5.19, we have taken $\theta = 10$ K, $L/h = 20$ with S–S support

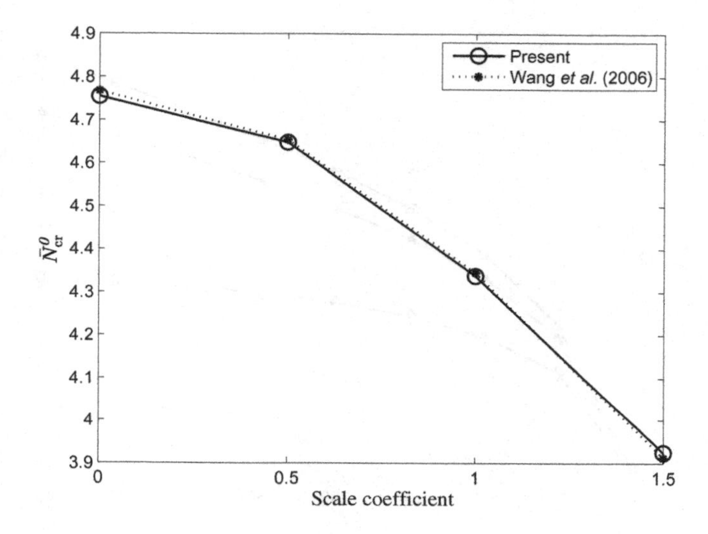

Fig. 5.17 Comparison of critical buckling load parameter (TBT)

Table 5.4 Comparison of critical buckling load parameter (RBT)

		S–S		C–C	
L/h	μ	Present	Ref.*	Present	Ref.*
10	0	9.6228	9.6228	35.8075	35.8075
	1	8.7583	8.7583	25.6724	25.6724
	2	8.0364	8.0364	20.0090	20.0090
	3	7.6149	7.4245	16.3927	16.3927

*Emam (2013).

and in Fig. 5.20, we have taken $\theta = 10K$, $L/h = 30$ with C–C support. In these figures, results have been shown for various values of scale coefficients. The Winkler modulus parameter is taken in the range of 0–400. It is observed from these figures that critical buckling load parameter ($\bar{N}^0_{cr}$) decreases with increase in scale coefficient. It may be noted that results associated with $e_0a = 0$ nm correspond to those of local beam theory. One may observe that the results obtained by local beam theory are overpredicted than that obtained by nonlocal beam theory. Therefore, nonlocal theory should be considered for buckling analysis of structures at nanoscale. It is seen that

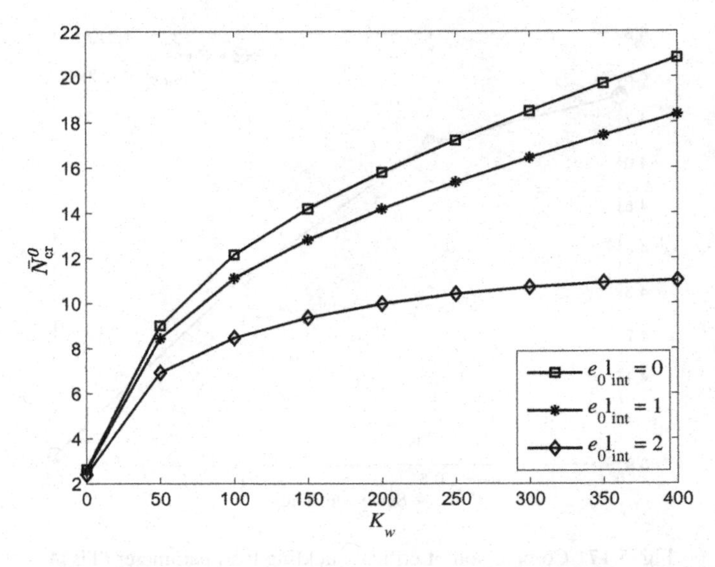

Fig. 5.18 Effect of the Winkler modulus parameter on critical buckling load parameter (EBT)

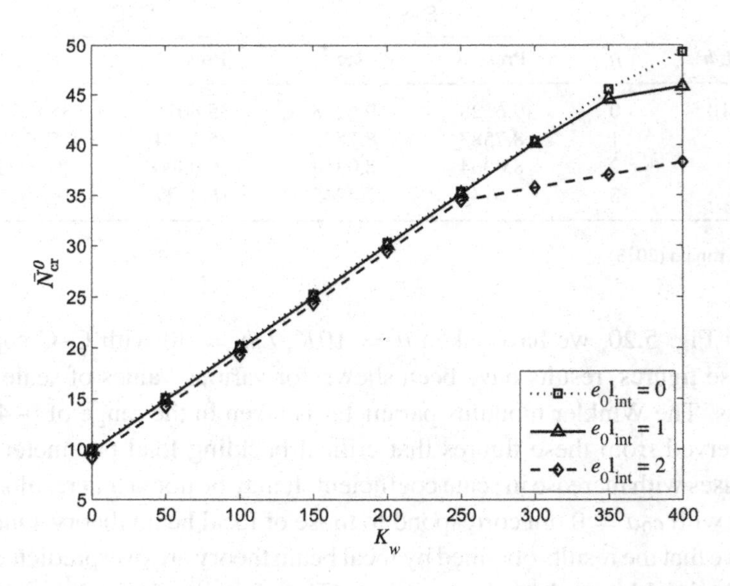

Fig. 5.19 Effect of the Winkler modulus parameter on critical buckling load parameter (TBT)

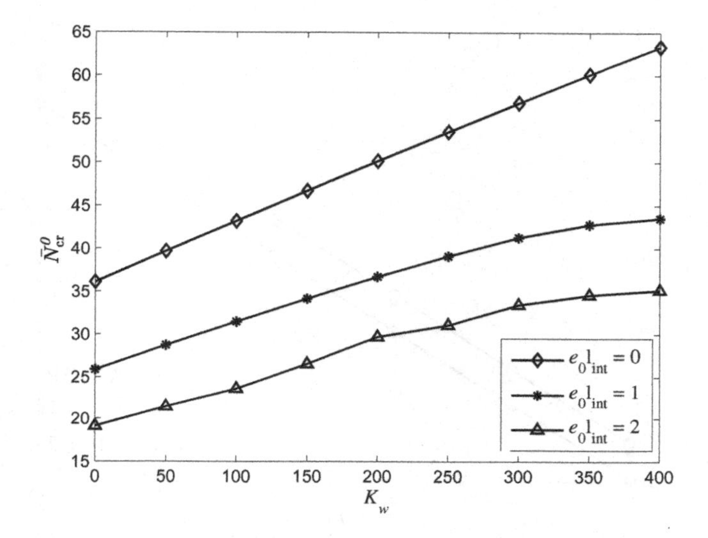

Fig. 5.20 Effect of the Winkler modulus parameter on critical buckling load parameter (RBT)

critical buckling load increases with increase in Winkler modulus parameter. This is because that the nanotube becomes stiffer when elastic medium constant is increased. In addition, it is also observed that critical buckling loads show nonlinear behavior with respect to stiffness of surrounding matrix for higher e_0a values. This may be due to the fact that increase of the Winkler modulus causes CNT to be more rigid.

5.2.4 *Effect of Pasternak shear modulus parameter*

In this section, effect of Pasternak shear modulus parameter on the buckling has been examined. As such, Figs. 5.21–5.23 show the distribution of critical buckling load parameter against Pasternak shear modulus for EBT, TBT, and RBT, respectively, in low-temperature environment. Numerical values of parameters are chosen as $K_w = 0, \theta = 10$ K, $L/h = 40$ with C–S support in Fig. 5.21, whereas in Fig. 5.22, we have taken $K_w = 50, \theta = 10$ K, $L/h = 20$ with C–S support and in Fig. 5.23, we have taken $K_w = 0$, $\theta = 10$ K, $L/h = 20$ with S–S edge condition. Graph is plotted for various values of scale coefficients with Pasternak shear modulus parameter ranging from 0 to 10. It is observed from the figures that critical buckling load

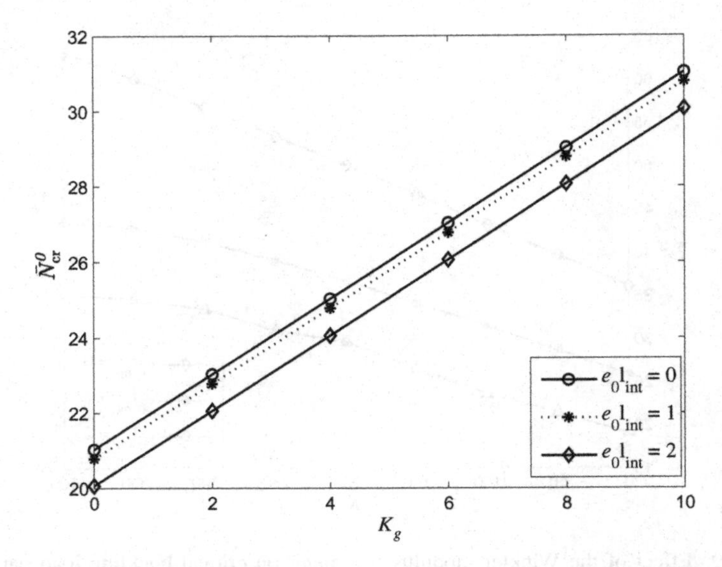

Fig. 5.21 Effect of Pasternak shear modulus parameter on critical buckling load parameter (EBT)

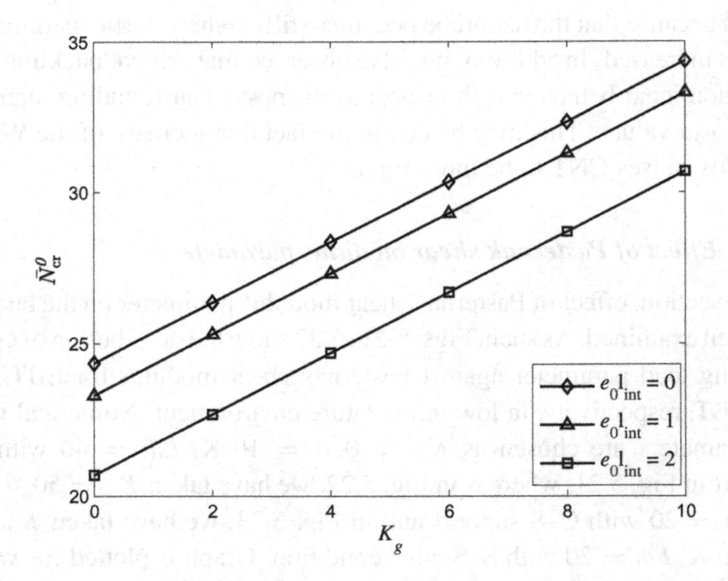

Fig. 5.22 Effect of Pasternak shear modulus parameter on critical buckling load parameter (TBT)

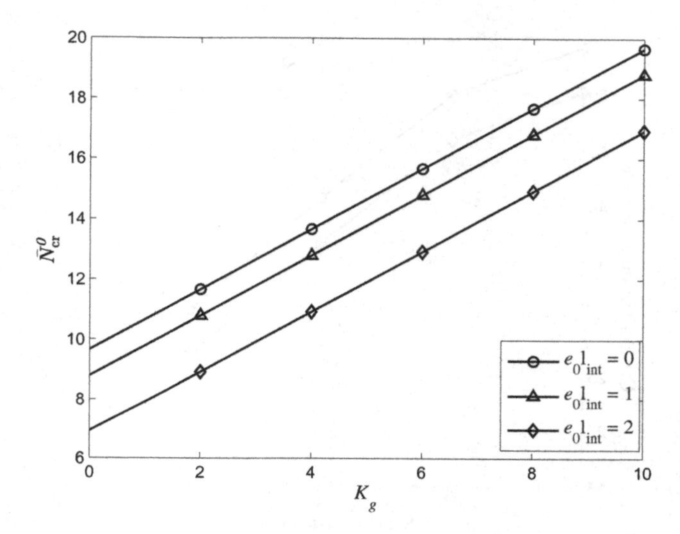

Fig. 5.23 Effect of Pasternak shear modulus parameter on critical buckling load parameter (RBT)

parameter associated increases with Pasternak shear modulus parameter. This increase is influenced by small scale coefficient. With increase in scale coefficient, critical buckling load parameter for a particular Pasternak shear modulus parameter decreases. Here it is also observed that unlike Winkler foundation model, the increase of critical buckling load parameter with Pasternak foundation is linear in nature. This is due to the domination of elastic medium modeled as the Pasternak-type foundation model. Same observation has also been reported in Murmu and Pradhan (2009b). Next, we have analyzed the effect of Pasternak foundation model over Winkler foundation model. As such, Fig. 5.24 illustrates the critical buckling load parameter of RBT nanobeams as a function of small scale coefficient in low-temperature environment with $L/h = 10$ and C–C edge condition. It may be observed that critical buckling load parameter obtained from Pasternak foundation model is relatively larger than those obtained from the Winkler foundation model.

5.2.5 *Effect of temperature*

Here, effect of temperature on the buckling of nanobeams embedded in elastic medium has been investigated. As such, Figs. 5.25–5.27 show variation

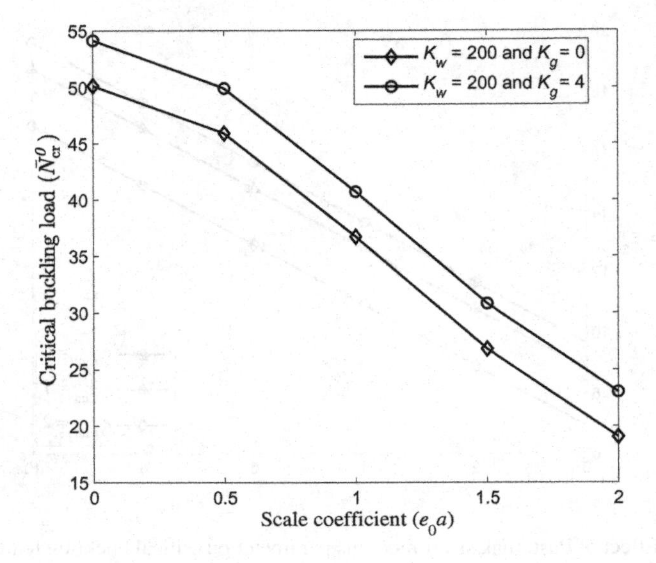

Fig. 5.24 Variation of $\bar{N}_{cr}^{0}$ with e_0a

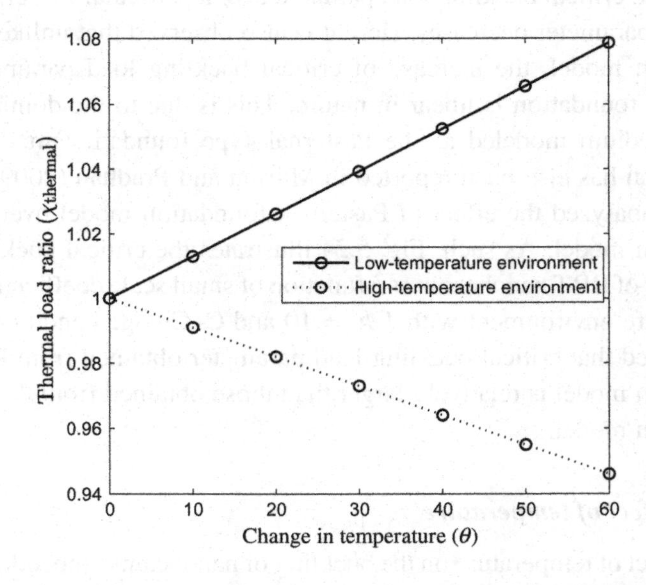

Fig. 5.25 Change in thermal load ratio with change in temperature (EBT)

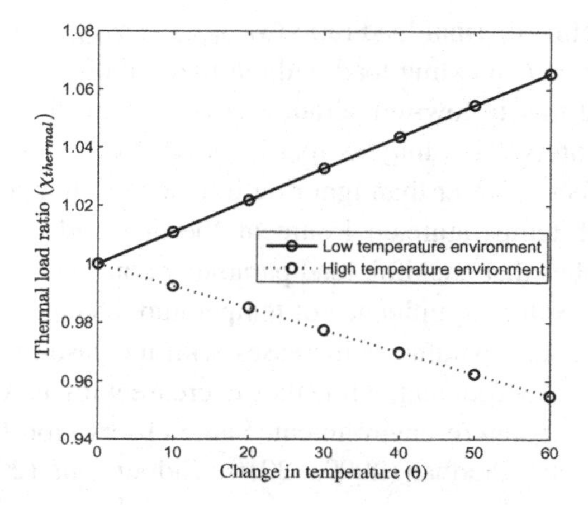

Fig. 5.26 Change in thermal load ratio with change in temperature (TBT)

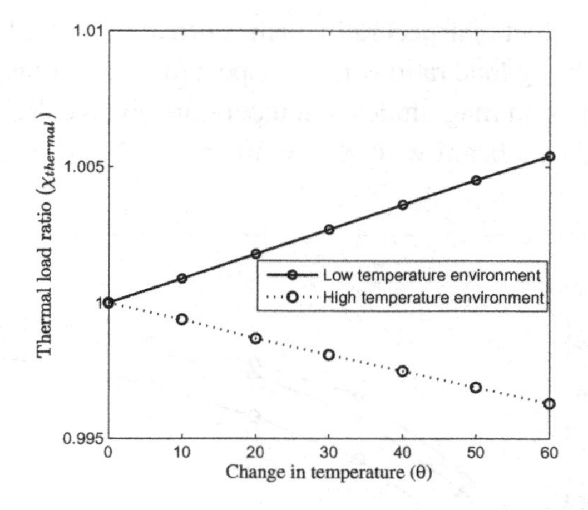

Fig. 5.27 Change in thermal load ratio with change in temperature (RBT)

of thermal load ratio (associated with first mode) with change in temperature (θ), respectively, for EBT, TBT, and RBT. In Fig. 5.25, we have taken S–S nanobeam with $L/h = 20$, $e_0a = 2$ nm, $K_w = 50$, $K_g = 2$. Similarly, we have taken $L/h = 50$, $e_0a = 0.5$ nm, $K_w = 50$, $K_g = 2$ with C–C support in Fig. 5.26 and $L/h = 10$, $e_0a = 1.5$ nm, $K_w = 50$, $K_g = 2$ with C–C support in Fig. 5.27.

Here, we define thermal load ratio (χ_{thermal}) as χ_{thermal} = Buckling load with thermal effect/ Buckling load without thermal effect.

It is noticed that in low-temperature environment, thermal load ratios are more than unity. This implies that buckling load parameter considering thermal effect is larger than ignoring influence of temperature change. Whereas in high-temperature environment, thermal load ratios are less than unity. This implies that buckling load parameter considering thermal effect is smaller than excluding influence of temperature change. In other words, critical buckling load parameter increases with increase in temperature in low-temperature environment, while they decrease with increase in temperature in high-temperature environment. Same observation have also been noted in Murmu and Pradhan (2009a, 2010), Zidour *et al.* (2012), Maachou *et al.* (2011).

5.2.6 *Effect of aspect ratio*

To illustrate the effect of aspect ratio on the critical buckling load parameter, variation of buckling load ratio with the aspect ratio (L/h) has been shown in Fig. 5.28 for different magnitudes of temperature change. Results have been shown for TBT nanobeam with $K_w = 50$, $K_g = 2$, $e_0a = 1$ nm, and S–S

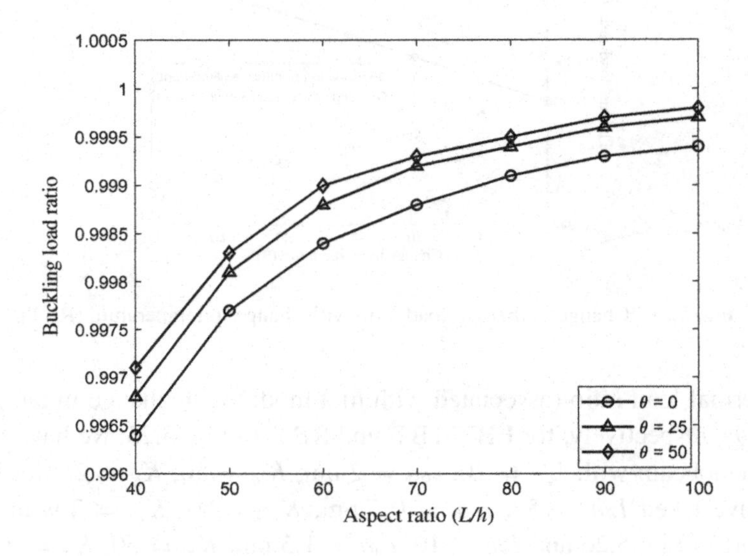

Fig. 5.28 Change in buckling load ratio with aspect ratio

edge condition. It is observed that buckling load ratio (associated with first mode) increases with increase in aspect ratio. In addition, it is also seen that critical load is also dependent on temperature change (θ). The differences in magnitudes of buckling load ratio for different temperature changes are larger in low aspect ratios while the differences in magnitudes of scale load ratio for different temperature changes are smaller for large aspect ratios. It is also seen that for larger temperature change, the rate of increase of buckling load ratio is less compared to smaller temperature change. One may note that same observation may also be seen in case of EBT and RBT beam theories.

5.3 Conclusions

DQM has been employed for buckling analysis of non-uniform nanobeams based on four different beam theories such as EBT, TBT, RBT, and LBT. Non-uniform material properties are assumed by taking exponentially varying stiffness. New results have been shown for two types of boundary conditions such as guided and simply supported-guided. Similarly, DQM has also been applied to investigate thermal effect on the buckling of nanobeams embedded in elastic medium based on nonlocal RBT. Theoretical formulations include effects of small scale, elastic medium, and temperature change. It is seen that results obtained based on local beam theory are overestimated. Critical buckling load parameters increase with increase in temperature and also with Winkler and Pasternak coefficients of elastic foundation.

Chapter 6

Vibration of Nanobeams

In this chapter, Rayleigh–Ritz method has been applied to study vibration of nanobeams based on Euler–Bernoulli beam theory (EBT) and Timoshenko beam theory (TBT). Next, differential quadrature method (DQM) has been used to investigate vibration of nanobeams based on four types of beams such as EBT, TBT, Reddy–Bickford beam theory (RBT), and Levinson beam theory (LBT). In the Rayleigh–Ritz method, both one-dimensional simple polynomials and boundary characteristic orthogonal polynomials have been used as shape functions.

6.1 Vibration of Nanobeams using Rayleigh–Ritz Method

In this analysis, we have used the following non-dimensional terms:

$$X = \frac{x}{L},$$

$$W = \frac{w_0}{L},$$

$$\alpha = \frac{e_0 a}{L} = \text{scaling effect parameter},$$

$$\xi = \frac{L\sqrt{A}}{I},$$

$$\tau = \frac{1}{\xi^2},$$

$$\lambda^2 = \frac{\rho A \omega^2 L^4}{EI} = \text{frequency parameter},$$

$$\Omega = \frac{EI}{k_s GAL^2} = \text{shear deformation parameter}.$$

Application of Rayleigh–Ritz method would convert vibration problems to generalized eigenvalue problems which have been discussed in Section 2.1.2. In generalized eigenvalue equation (Eq. (3.21)), stiffness and mass matrices for EBT are given by

$$K(i,j) = \int_0^1 \varphi_i''\varphi_j''\,dX,$$

$$M_a(i,j) = \int_0^1 \varphi_i\varphi_j - \frac{\alpha^2}{2}\varphi_i\varphi_j'' - \frac{\alpha^2}{2}\varphi_i''\varphi_j\,dX.$$

Similarly, in the generalized eigenvalue equation for TBT nanobeams (Eq. (3.26)), matrices $[K]$ and $[M_a]$ defined are as

$$K = \begin{bmatrix} k_1 & k_2 \\ k_3 & k_4 \end{bmatrix},$$

where k_1, k_2, k_3, and k_4 are submatrices and are given by

$$k_1(i,j) = \int_0^1 \varphi_i'\varphi_j\,dX,$$

$$k_2(i,j) = \int_0^1 \varphi_i'\psi_j\,dX,$$

$$k_3(i,j) = \int_0^1 \psi_i\varphi_j'\,dX,$$

$$k_4(i,j) = \int_0^1 (\psi_i\psi_j + \Omega\psi_i'\psi_j')\,dX,$$

$$M_a = \begin{bmatrix} m_1 & m_2 \\ m_3 & m_4 \end{bmatrix},$$

where m_1, m_2, m_3, and m_4 are submatrices and are given as

$$m_1(i,j) = \Omega\int_0^1 \varphi_i\varphi_j\,dX,$$

$$m_2(i,j) = \Omega\frac{\alpha^2}{2}\int_0^1 \varphi_i\psi_j'\,dX,$$

$$m_3(i,j) = \Omega\frac{\alpha^2}{2}\int_0^1 \psi_i'\varphi_j\,dX,$$

$$m_4(i,j) = \Omega\int_0^1 (\tau\psi_i\psi_j + \tau\alpha^2\psi_i'\psi_j')\,dX.$$

6.2 Numerical Results and Discussions

Frequency parameters of single-walled carbon nanotubes (SWCNTs) have been computed by using Rayleigh–Ritz method. In the numerical evaluations, following parameters of SWCNTs have been used (Wang *et al.* 2007): diameter, $d = 0.678$ nm, $L = 10d$, $t = 0.066$, $k_s = 0.563$, $E = 5.5$ TPa, $G = E/[2(1 + v)]$, $\nu = 0.19$, and $I = \Pi d^4/64$.

In this study, frequency parameters of both Euler–Bernoulli and Timoshenko nanobeams have been computed. Results have been investigated for different scaling effect parameters and boundary conditions. Firstly, frequency parameters are being computed taking simple polynomials of the form X^{i-1} in the Rayleigh–Ritz method. Then the polynomials are orthogonalized by Gram–Schmidt process and are used in the Rayleigh–Ritz method to obtain frequency parameters. Table 6.1 shows the convergence studies of first three frequency parameters ($\sqrt{\lambda}$) for S–S and C–S Euler–Bernoulli nanobeams taking $\alpha = 0.5$ and $L = 10d$. Similarly, convergence studies of first three frequency parameters for S–S and C–S Timoshenko nanobeams are tabulated in Table 6.2 for $\alpha = 0.5$ and $L = 10d$. In these tables, it is observed that the frequency parameters are close to the results of Wang *et al.* (2007) as the value of n increases. In Table 6.3, the first four frequency parameters of Euler–Bernoulli nanobeams have been compared with results of Wang *et al.* (2007) and are found to be in good agreement. Similarly, the results of Timoshenko nanobeams subjected to various boundary conditions

Table 6.1 Convergence of the first three frequency parameters for Euler–Bernoulli nanobeams

	S–S			C–S		
n	First	Second	Third	First	Second	Third
3	2.3026	3.8475	5.0587	2.7928	3.9140	5.6488
4	2.3026	3.4688	5.0587	2.7900	3.8530	4.8090
5	2.3022	3.4688	4.3231	2.7899	3.8341	4.6708
6	2.3022	3.4604	4.3231	2.7899	3.8327	4.6194
7	2.3022	3.4604	4.2945	2.7899	3.8325	4.6122
8	2.3022	3.4604	4.2945	2.7899	3.8325	4.6106
9	2.3022	3.4604	4.2941	2.7899	3.8325	4.6105
10	2.3022	3.4604	4.2941	2.7899	3.8325	4.6105
11	2.3022	3.4604	4.2941	2.7899	3.8325	4.6105

Table 6.2 Convergence of the first three frequency parameters for Timoshenko nanobeams

	S–S			C–S		
n	First	Second	Third	First	Second	Third
3	2.3867	3.6631	10.4677	2.7315	4.1148	6.8252
4	2.2760	3.6630	4.5482	2.7210	3.6916	4.8857
5	2.2760	3.3477	4.5481	2.7186	3.6521	4.3489
6	2.2756	3.3477	4.0425	2.7186	3.6373	4.2753
7	2.2756	3.3423	4.0425	2.7186	3.6364	4.2391
8	2.2756	3.3426	4.0212	2.7186	3.6362	4.2352
9	2.2756	3.3423	4.0212	2.7186	3.6362	4.2341
10	2.2756	3.3423	4.0209	2.7186	3.6362	4.2341
11	2.2756	3.3423	4.0209	2.7186	3.6362	4.2341

Table 6.3 Validation of the first four frequency parameters of Euler–Bernoulli nanobeams

	$\alpha = 0$		$\alpha = 0.1$		$\alpha = 0.3$	
Mode No.	Present	Ref.*	Present	Ref.*	Present	Ref.*
S–S						
1	3.1416	3.1416	3.0685	3.0685	2.6800	2.6800
2	6.2832	6.2832	5.7817	5.7817	4.3013	4.3013
3	9.4248	9.4248	8.0400	8.0400	5.4423	5.4422
4	12.566	12.566	9.9162	9.9162	6.3630	6.3630
C–S						
1	3.9266	3.9266	3.8209	3.8209	3.2828	3.2828
2	7.0686	7.0686	6.4649	6.4649	4.7668	4.7668
3	10.210	10.210	8.6517	8.6517	5.8371	5.8371
4	13.252	13.352	10.469	10.469	6.7145	6.7143
C–C						
1	4.7300	4.7300	4.5945	4.5945	3.9184	3.9184
2	7.8532	7.8532	7.1402	7.1402	5.1963	5.1963
3	10.996	10.996	9.2583	9.2583	6.2317	6.2317
4	14.137	14.137	11.016	11.016	7.0482	7.0482

*Wang *et al.* (2007).

Table 6.4 Validation of the first four frequency parameters of Timoshenko nanobeams

	$\alpha = 0$		$\alpha = 0.1$		$\alpha = 0.3$	
Mode No.	Present	Ref.*	Present	Ref.*	Present	Ref.*
S–S						
1	3.0742	3.0929	3.0072	3.0243	2.6412	2.6538
2	5.9274	5.9399	5.4400	5.5304	4.1357	4.2058
3	8.4057	8.4444	7.3662	7.4699	5.0744	5.2444
4	10.601	10.626	8.9490	8.9874	6.0173	6.0228
C–S						
1	3.7336	3.7845	3.6476	3.6939	3.1784	3.2115
2	6.2945	6.4728	6.0015	6.0348	4.4926	4.6013
3	8.4762	8.1212	7.5816	7.8456	5.3307	5.5482
4	10.361	10.880	9.2044	9.2751	6.2286	6.2641
C–C						
1	4.3980	4.4491	4.3026	4.3471	3.7578	3.7895
2	6.7711	6.9524	6.3507	6.4952	4.8196	4.9428
3	9.1185	9.1626	8.1274	8.1969	5.6082	5.8460
4	11.014	11.113	9.1456	9.5447	6.1194	6.4762

*Wang *et al.* (2007).

have been compared with Wang *et al.* (2007) in Table 6.4 for different scaling effect parameters. From Tables 6.3 and 6.4, it can be clearly seen that the nonlocal results are smaller than the corresponding local ones. Frequency parameters of nanobeams subjected to F–F and S–F boundary conditions have been given in Table 6.5 for different scaling effect parameters. It may be noted that the frequency parameters obtained by using orthonormalized polynomials are same as that of using simple polynomials. But here the computations become more efficient and less time is required for the execution of the program. It is due to the fact that some of the matrix elements become zero or one due to the orthonormality. One of the interesting facts in this analysis is that C–C nanobeams have the highest frequency parameters than other boundary conditions. It helps the design engineers to obtain desired frequency parameters as per the application.

The behavior of scaling effect parameter on the frequency parameter is shown in Figs. 6.1–6.3, respectively, for S–S, C–S, and C–C Euler–Bernoulli nanobeams. Similarly, Figs. 6.4–6.6 show variation of

Table 6.5 First four frequency parameters of Timoshenko nanobeams and some new boundary conditions

Mode No.	$\alpha = 0$	$\alpha = 0.1$	$\alpha = 0.3$	$\alpha = 0.5$	$\alpha = 0.7$
S–S					
1	0.0009	0.0008	0.0001	0.0001	0.0001
2	3.8065	3.7118	3.2121	2.7378	2.3931
3	6.4684	6.0146	4.5340	3.6575	3.1302
4	8.7295	7.7276	5.3708	4.2542	3.6193
F–F					
1	0.0009	0.0008	0.0008	0.0004	0.0004
2	4.5443	4.4253	3.8029	3.2201	2.8043
3	7.0857	6.5603	4.8810	3.9150	3.3428
4	9.2673	8.1717	5.6529	4.4805	3.8132

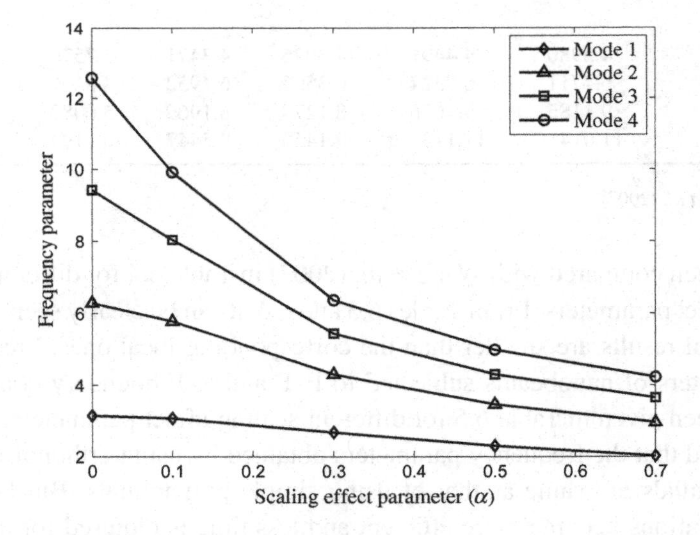

Fig. 6.1 Change in frequency parameter with α (S–S)

the frequency parameter with the scaling effect parameter, respectively, for S–S, C–S, and C–C Timoshenko nanobeams. In these figures, the first four frequency parameters have been shown for both Euler–Bernoulli and Timoshenko nanobeams. From these figures, it is depicted that frequency parameters are overpredicted when local beam model is considered for vibration analysis of nanobeams. As the scaling effect parameter increases,

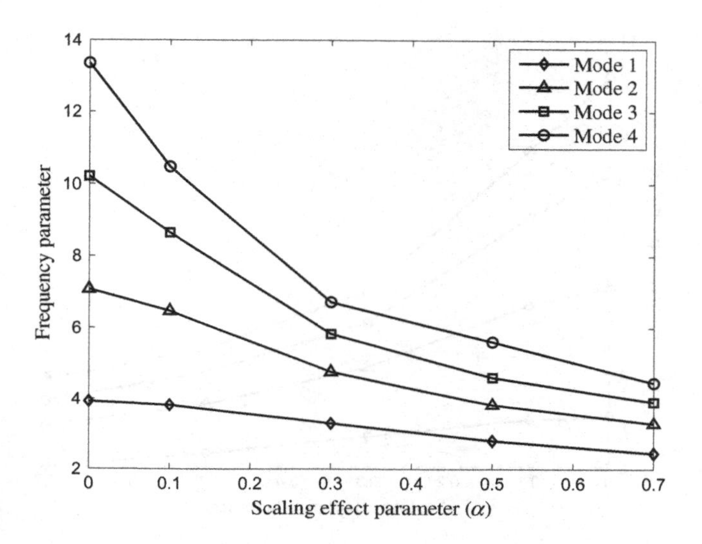

Fig. 6.2 Change in frequency parameter with α (C–S)

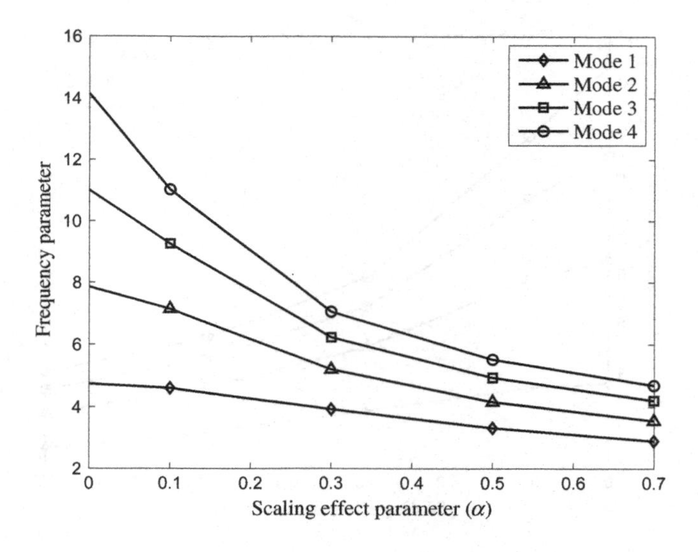

Fig. 6.3 Change in frequency parameter with α (C–C)

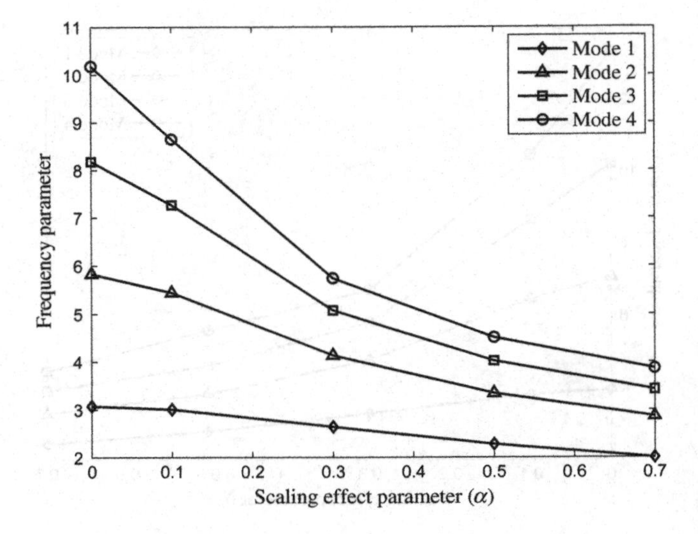

Fig. 6.4 Change in frequency parameter with α (S–S)

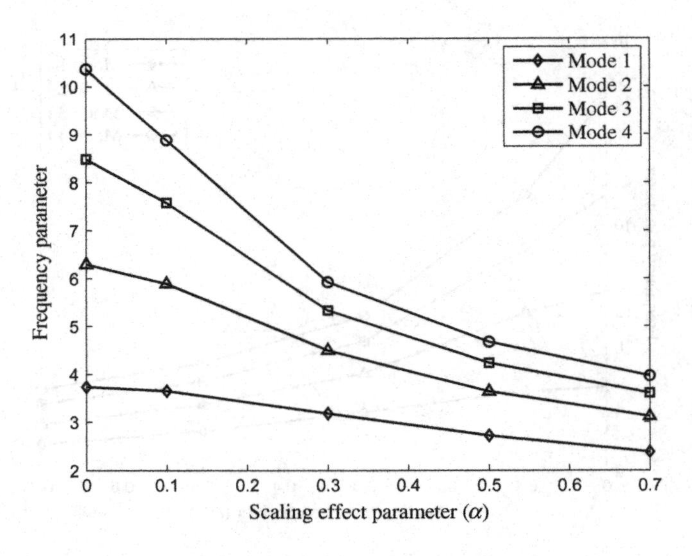

Fig. 6.5 Change in frequency parameter with α (C–S)

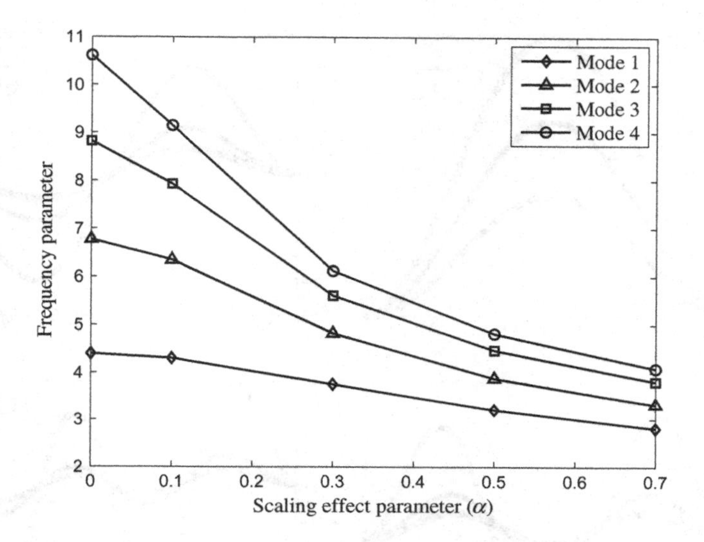

Fig. 6.6 Change in frequency parameter with α (C–C)

frequency parameters of nonlocal nanobeams become smaller than those of its local counterpart. This reduction can be clearly seen when we consider higher vibration modes. The reduction is due to the fact that the nonlocal model may be viewed as atoms linked by elastic springs, while in the case of local continuum model, the spring constant is assumed to take an infinite value. So small scale effect makes the nanobeams more flexible and nonlocal impact cannot be neglected. As such, nonlocal theory should be used for better predictions of high natural frequency of nanobeams. Mode shapes are useful for engineers to design structures because they represent the shape that the structures will vibrate in free motion. Sometimes, the knowledge of higher modes is necessary before finalizing the design of an engineering system. Thus, while studying vibration problems, viz. beam, plate or shell, one may always see the tabulation of the higher frequencies in the open literature. As such, the present investigators have reported first few higher modes in Fig. 6.7 for benchmarking the results which may help the researchers of nanotechnology. In Fig. 6.7, we have given the first four deflections of nonlocal C–C Euler–Bernoulli nanobeams with scaling effect parameters as 0, 0.3, and 0.6. It can be seen that mode shapes are affected

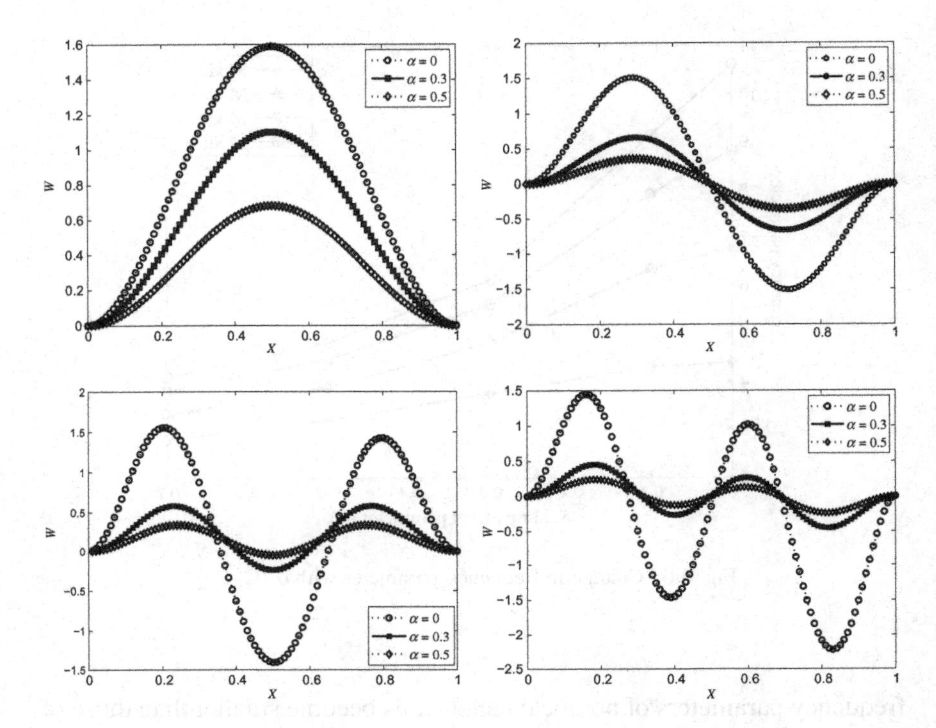

Fig. 6.7 First four deflection shapes of c–c nanobeams

by small length scale. By understanding the modes of vibration, we can better design the structures as per the need.

6.2.1 *Vibration of nanobeams using DQM*

In this section, DQM has been employed to study various nonlocal beam theories such as EBT, TBT, RBT, and LBT. Boundary conditions have been substituted in the coefficient matrices.

In this problem, we have introduced the following non-dimensional terms:

$$X = \frac{x}{L},$$

$$W = \frac{w_0}{L},$$

$$\alpha = \frac{e_0 a}{L} = \text{scaling effect parameter},$$

$$\xi = \frac{L\sqrt{A}}{I},$$

$$\tau = \frac{1}{\xi^2},$$

$$\lambda^2 = \frac{\rho A \omega^2 L^4}{EI} = \text{frequency parameter},$$

$$\Omega = \frac{EI}{K_s GAL^2} = \text{shear deformation parameter},$$

$$\bar{\Omega} = \frac{G\tilde{A}L^2}{EI},$$

$$\hat{\Omega} = \frac{EI}{G\bar{A}L^2}.$$

Below, we have included the non-dimensionalized forms of the governing differential equations, respectively, for EBT, TBT, RBT, and LBT in Eqs. (6.1)–(6.4).

$$\frac{d^4 W}{dX^4} = \lambda^2 \left(W - \alpha^2 \frac{d^2 W}{dX^2} \right), \tag{6.1}$$

$$\frac{d^4 W}{dX^4} = \lambda^2 \left(-\tau \frac{d^2 W}{dX^2} + \alpha^2 \tau \frac{d^4 W}{dX^4} - \Omega \frac{d^2 W}{dX^2} + \Omega \alpha^2 \frac{d^4 W}{dX^4} + W - \alpha^2 \frac{d^2 W}{dX^2} \right), \tag{6.2}$$

$$\bar{\Omega} \frac{5}{4} \frac{d^4 W}{dx^4} - \frac{1}{105} \frac{d^6 W}{dx^6} = \lambda^2 \left(-\frac{17}{21} \frac{d^2 W}{dX^2} + \frac{17}{21} \alpha^2 \frac{d^4 W}{dX^4} + \frac{105}{84} \bar{\Omega} \left(W - \alpha^2 \frac{d^2 W}{dX^2} \right) \right), \tag{6.3}$$

$$\frac{d^4 W}{dX^4} = \lambda^2 \left[W + \left(\alpha^2 \tau + \frac{4}{5} \alpha^2 \hat{\Omega} \right) \frac{d^4 W}{dX^4} - \left(\tau + \alpha^2 + \frac{4}{5} \hat{\Omega} \right) \frac{d^2 W}{dX^2} \right]. \tag{6.4}$$

By the application of DQM, one may obtain generalized eigenvalue problem as

$$[S]\{W\} = \lambda^2 [M_a]\{W\}, \tag{6.5}$$

where S is the stiffness matrix and M_a is the mass matrix.

6.3 Numerical Results and Discussions

Frequency parameters ($\sqrt{\lambda}$) have been obtained by solving Eq. (6.5) using computer code developed by the authors. The lowest four eigenvalues corresponding to the first four frequency parameters have been reported for different boundary conditions. In this investigation, various parameters used for numerical evaluations are as follows (Reddy 2007): $E = 30 \times 10^6$, $\nu = 0.3$, $L = 10$ nm; $G = E/2(1+\nu)$, $k_s = \frac{5}{6}$, and unless mentioned $L/h = 10$.

6.3.1 *Convergence*

A convergence study is being carried out for finding the minimum number of grid points to obtain the converged results. Lower frequency parameters converge first than successive higher frequency parameters. The first three frequency parameters converge with less number of grid points than fourth one. Hence, the convergence of fourth frequency parameter ($\sqrt{\lambda_4}$) of EBT and RBT nanobeams is shown in Fig. 6.8. In this graph, μ is taken as 1 nm^2 with S–S boundary condition. It is seen that the fourth frequency parameter converges at 16 grid points. Hence, 16 grid points are taken for obtaining the converged results of the first four frequency parameters of nanobeams based on four beam theories.

6.3.2 *Validation*

Validation of the proposed method is confirmed by comparing the obtained results with those available in literature (Wang *et al.* 2007; Reddy 2007). For this purpose, same parameters as used in Wang *et al.* (2007) and Reddy (2007) are taken. Comparison of the fundamental frequency parameter has been shown in Table 6.6 for S–S edge condition. Results have been compared for all four types of beam theories. In this table, μ is taken from 0 to 4 nm^2. Similarly, comparison of the results of Timoshenko nanobeams

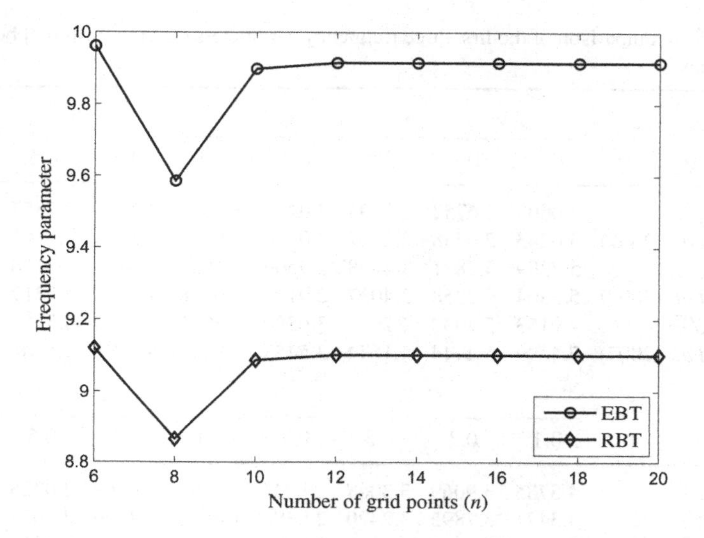

Fig. 6.8 Variation of frequency parameter with grid points

Table 6.6 Comparison of fundamental frequency parameter (λ) for S–S nanobeams

μ	EBT	Ref.*	TBT	Ref.*	RBT	Ref.*	LBT	Ref.*
0	9.8696	9.8696	9.7000	9.7454	9.7000	9.7454	9.7000	9.7657
0.5	9.6347	9.6347	9.4953	9.5135	9.5100	9.5135	9.4755	9.5333
1	9.4159	9.4159	9.2973	9.2973	9.3000	9.2974	9.2603	9.3168
1.5	9.2113	9.2113	9.1113	9.0953	9.1000	9.0954	9.0591	9.1144
2	9.0195	9.0195	8.9359	8.9059	8.9100	8.9060	8.8704	8.9246
2.5	8.8392	8.8392	8.7703	8.7279	8.7300	8.7279	8.6931	8.7462
3.0	8.6693	8.6693	8.6136	8.5601	8.5600	8.5602	8.5260	8.5780
3.5	8.5088	8.5088	8.4650	8.4017	8.4017	8.4017	8.3682	8.4193
4.0	8.3569	8.3569	8.3238	8.2517	8.2517	8.2517	8.2188	8.2690

*Reddy (2007).

subjected to different boundary conditions (S–S, C–C, C–S, C–F) has been reported in Table 6.7 for different scaling effect parameters. It is observed from Table 6.6 that the fundamental frequency parameter of EBT nanobeams is higher than other types of nanobeams. One may see from Table 6.7 that no non-trivial real frequencies will exist for cantilever beams once the scaling effect parameter approaches the value 0.6138. This is due to the

Table 6.7 Comparison of the first three frequency parameters ($\sqrt{\lambda}$) for different boundary conditions

	S–S				C–S			
α	0.1	0.3	0.5	0.7	0.1	0.3	0.5	0.7
$\sqrt{\lambda_1}$	3.0603	2.6752	2.2994	2.0193	3.8089	3.2754	2.7857	2.4328
Wang *et al.* (2007)	3.0243	2.6538	2.2867	2.0106	3.6939	3.2115	2.7471	2.4059
$\sqrt{\lambda_2}$	5.7309	4.2831	3.4498	2.9506	6.4030	4.7424	3.8170	3.2654
Wang *et al.* (2007)	5.5304	4.2058	3.4037	2.9159	6.0348	4.6013	3.7312	3.2003
$\sqrt{\lambda_3}$	7.9153	5.4033	4.2693	3.6290	8.5055	5.7870	4.5767	3.8929
Wang *et al.* (2007)	7.4699	5.2444	4.1644	3.5453	7.8456	5.5482	4.4185	3.7666
	C–C				C–F			
	0.1	0.3	0.5	0.7	0.1	0.3	0.5	0.7
$\sqrt{\lambda_1}$	4.5785	3.9069	3.3068	2.8823	1.8796	1.9158	2.0225	—
Wang *et al.* (2007)	4.3471	3.7895	3.2420	2.8383	1.8650	1.8999	2.0024	—
$\sqrt{\lambda_2}$	7.0405	5.1223	4.0730	3.4473	4.5303	3.7543	2.9355	—
Wang *et al.* (2007)	6.4952	4.9428	3.9940	3.4192	4.3506	3.6594	2.8903	—
$\sqrt{\lambda_3}$	9.0856	6.1696	4.8909	4.1600	7.0710	5.2652	—	—
Wang *et al.* (2007)	8.1969	5.8460	4.6769	3.9961	6.6091	5.0762	—	—

fact that successive odd and even vibration modes approach each other and are suppressed with the increasing value of α. In addition, real frequencies can be obtained for only first few modes once value of α approaches 0.6138. Similar findings are also reported in Wang *et al.* (2007) and Reddy (2007). We have also incorporated graphical comparison in Fig. 6.9 with Wang *et al.* (2007) for TBT nanobeams with S–S support. Similarly, Fig. 6.10 shows the graphical comparison with Reddy (2007) for Reddy nanobeams with $L/h = 10$. One may observe that a close agreement of the results is achieved.

6.3.3 *Effect of nonlocal parameter*

Effect of nonlocal parameter on the first four frequency parameters ($\sqrt{\lambda}$) of nanobeams based on four beam theories is analyzed. In this analysis, boundary conditions such as S–S, C–S, C–C, and C–F are taken into consideration. Both tabular and graphical results are presented in this context. Table 6.8 shows the first four frequency parameters of S–S, C–S, C–C, and C–F EBT nanobeams for different nonlocal parameters. From this table, it is seen that frequency parameters decrease with increase

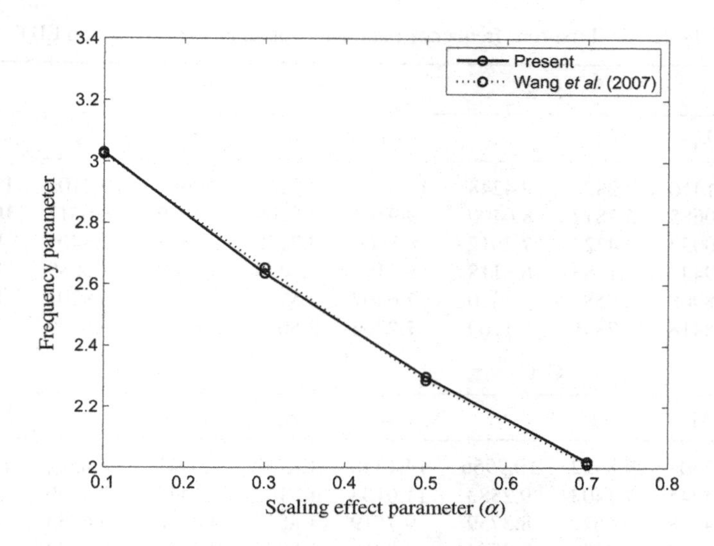

Fig. 6.9 Comparison of results with Wang *et al.* (2007)

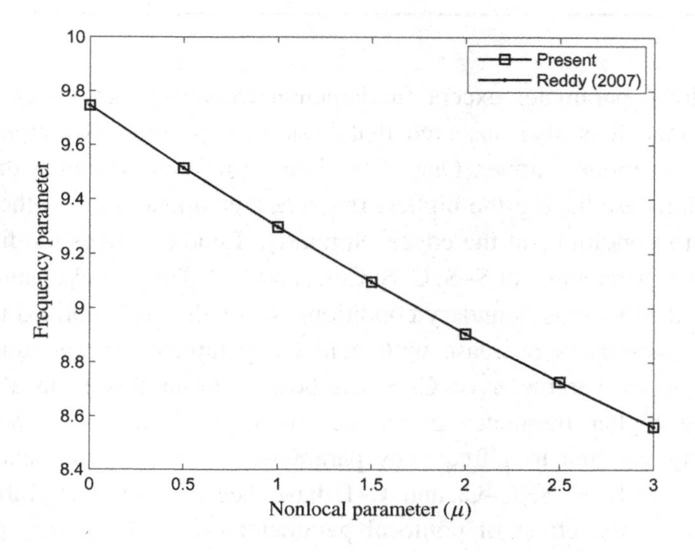

Fig. 6.10 Comparison of results with Reddy (2007)

Table 6.8 First four frequency parameters of nanobeams based on EBT

	S–S				C–S			
μ	λ_1	λ_2	λ_3	λ_4	λ_1	λ_2	λ_3	λ_4
0	3.1416	6.2832	9.4248	12.5664	3.9266	7.0686	10.2102	13.3518
1	3.0685	5.7817	8.0400	9.9161	3.8209	6.4649	8.6517	10.4688
2	3.0032	5.4324	7.3012	8.8000	3.7278	6.0545	7.8405	9.2811
3	2.9444	5.1683	6.8118	8.1195	3.6448	5.7488	7.3089	8.5619
4	2.8908	4.9581	6.4520	7.6407	3.5701	5.5079	6.9204	8.0573
5	2.8418	4.7846	6.1709	7.2764	3.5024	5.3107	6.6179	7.6740
	C–C				C–F			
	λ_1	λ_2	λ_3	λ_4	λ_1	λ_2	λ_3	λ_4
0	4.7300	7.8532	10.9956	14.1358	1.8751	4.6941	7.8548	10.9955
1	4.5945	7.1403	9.2583	11.0138	1.8792	4.5475	7.1459	9.2569
2	4.4758	6.6629	8.3739	9.7519	1.8833	4.4170	6.6753	8.3683
3	4.3707	6.3108	7.8004	8.9916	1.8876	4.2994	6.3322	7.7877
4	4.2766	6.0352	7.3840	8.4593	1.8919	4.1924	6.0674	7.3617
5	4.1917	5.8107	7.0611	8.0551	1.8964	4.0942	5.8550	7.0272

in nonlocal parameter except fundamental frequency parameter of C–F nanobeams. It is also observed that frequency parameters increase with increase in mode number. One of the interesting observation is that C–C nanobeams are having the highest frequency parameters than other set of boundary conditions at the edges. Similarly, Table 6.9 gives the first four frequency parameters of S–S, C–S, C–C, and C–F Timoshenko nanobeams subjected to various boundary conditions. Here also, it is noticed that frequency parameters decrease with nonlocal parameter except fundamental frequency parameter of C–F nanobeams. From this table also, one may see higher frequency parameters in the case of C–C nanobeams. Similarly, the first four frequency parameters of Reddy nanobeams subjected to S–S, C–S, C–C, and C–F have been reported in Table 6.10 to illustrate the effect of nonlocal parameter on the frequency parameters. From this table, one may conclude same observation as that of EBT and RBT nanobeams. Frequency parameters of Levison nanobeams subjected to different boundary conditions have been presented in Table 6.11. Here, one may observe that frequency parameters decrease with increase in nonlocal parameter except fundamental frequency parameter of C–F

Table 6.9 First four frequency parameters of nanobeams based on TBT

μ	S–S λ_1	S–S λ_2	S–S λ_3	S–S λ_4	C–S λ_1	C–S λ_2	C–S λ_3	C–S λ_4
0	3.1155	6.0867	8.8180	11.2766	3.8887	6.8298	9.5203	11.9354
1	3.0492	5.6421	7.6300	9.0990	3.7905	6.2794	8.1477	9.5019
2	2.9893	5.3236	6.9697	8.1207	3.7033	5.8978	7.4094	8.4532
3	2.9349	5.0786	6.5211	7.5123	3.6252	5.6101	6.9187	7.8093
4	2.8851	4.8813	6.1876	7.0798	3.5545	5.3815	6.5574	7.3547
5	2.8393	4.7172	5.9251	6.7487	3.4902	5.1934	6.2747	7.0080
	C–C λ_1	C–C λ_2	C–C λ_3	C–C λ_4	C–F λ_1	C–F λ_2	C–F λ_3	C–F λ_4
0	4.6813	7.5696	10.2199	12.5894	1.8800	4.6400	7.5700	10.2200
1	4.5494	6.8946	8.6377	9.8772	1.8801	4.5001	6.9012	8.6358
2	4.4338	6.4408	7.8215	8.7567	1.8838	4.3743	6.4531	7.8164
3	4.3312	6.1048	7.2888	8.0772	1.8875	4.2611	6.1242	7.2788
4	4.2394	5.8412	6.9005	7.6002	1.8913	4.1584	5.8686	6.8844
5	4.1563	5.6261	6.5987	7.7007	1.8953	4.0643	5.6624	6.5754

Table 6.10 First four frequency parameters of nanobeams based on RBT

μ	S–S λ_1	S–S λ_2	S–S λ_3	S–S λ_4	C–S λ_1	C–S λ_2	C–S λ_3	C–S λ_4
0	3.1218	6.1317	8.9488	11.5349	3.8978	6.8844	9.6685	12.2178
1	3.0492	5.6423	7.6340	9.1022	3.7905	6.2797	8.1492	9.5063
2	2.9843	5.3015	6.9325	8.0777	3.6963	5.8728	7.3713	8.4116
3	2.9258	5.0437	6.4678	7.4531	3.6125	5.5714	6.8646	7.7523
4	2.8726	4.8385	6.1262	7.0135	3.5374	5.3347	6.4955	7.2910
5	2.8239	4.6693	5.8592	6.6791	3.4693	5.1413	6.2087	6.9413
	C–C λ_1	C–C λ_2	C–C λ_3	C–C λ_4	C–F λ_1	C–F λ_2	C–F λ_3	C–F λ_4
0	4.6930	7.6344	10.3861	12.8968	1.8800	4.6500	7.6400	10.3900
1	4.5494	6.8951	8.6398	9.8831	1.8801	4.5001	6.9017	8.6380
2	4.4250	6.4132	7.7834	8.7171	1.8841	4.3645	6.4261	7.7778
3	4.3156	6.0626	7.2350	8.0228	1.8882	4.2434	6.0835	7.2238
4	4.2183	5.7904	6.8394	7.5395	1.8924	4.1341	5.8205	6.8211
5	4.1309	5.5700	6.5338	7.1741	1.8968	4.0345	5.6102	6.5071

Table 6.11 First four frequency parameters of nanobeams based on LBT

	S–S				C–S			
μ	λ_1	λ_2	λ_3	λ_4	λ_1	λ_2	λ_3	λ_4
0	3.1155	6.0867	8.8180	11.2766	3.8887	6.8298	9.5203	11.9354
1	3.0431	5.6008	7.5224	8.8983	3.7810	6.2253	8.0141	9.2720
2	2.9783	5.2625	6.8312	7.8968	3.6864	5.8197	7.2459	8.2010
3	2.9199	5.0067	6.3732	7.2861	3.6025	5.5197	6.7461	7.5567
4	2.8669	4.8030	6.0366	6.8566	3.5272	5.2843	6.3824	7.1062
5	2.8183	4.6350	5.7736	6.5295	3.4590	5.0921	6.0999	6.7647
	C–C				C–F			
	λ_1	λ_2	λ_3	λ_4	λ_1	λ_2	λ_3	λ_4
0	4.6813	7.5696	10.2199	12.5894	1.8800	4.6400	7.5700	10.2200
1	4.5354	6.8242	8.4777	9.6161	1.8804	4.4853	6.8312	8.4757
2	4.4092	6.3417	7.6303	8.4753	1.8844	4.3482	6.3549	7.6247
3	4.2986	5.9919	7.0893	7.7974	1.8884	4.2261	6.0128	7.0784
4	4.2004	5.7211	6.6996	7.3262	1.8926	4.1162	5.7507	6.6821
5	4.1123	5.5020	6.3989	6.9701	1.8969	4.0161	5.5412	6.3737

nanobeams. Next, to highlight the importance of nonlocal theory, variation of frequency ratio (frequency parameter calculated using nonlocal theory/ frequency parameter calculated using local theory) associated with the first four mode numbers with (e_0a) has been shown in Figs. 6.11–6.14, respectively, for EBT, TBT, RBT, and LBT. In these figures, we have taken S–S boundary condition. This frequency ratio serves as an index to estimate quantitatively the small scale effect on the vibration solution. It is clearly seen from the figures that the frequency ratios are less than unity. This implies that application of local beam model for vibration analysis of carbon nanotubes would lead to overprediction of the frequency in particular higher frequency if the small length scale effect between individual carbon atoms is neglected. Hence, the nonlocal beam theory should be used for better predictions of higher frequencies of nanobeams.

One of the important observation seen in this analysis is that frequency parameters of TBT, RBT, and LBT nanobeams are having approximately the same results. EBT nanobeam overpredicts frequency parameters than TBT, RBT, and LBT nanobeams since it neglects transverse shear and stain. In all the beam theories, fundamental frequency parameter decreases with

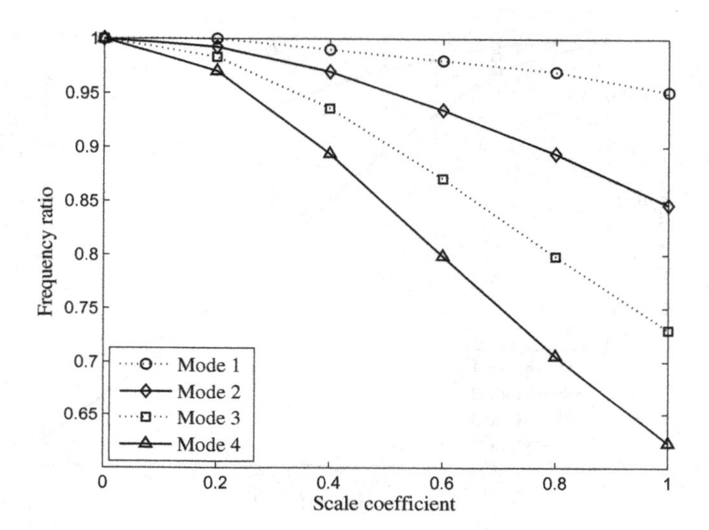

Fig. 6.11 Variation of frequency ratio with e_0a (EBT)

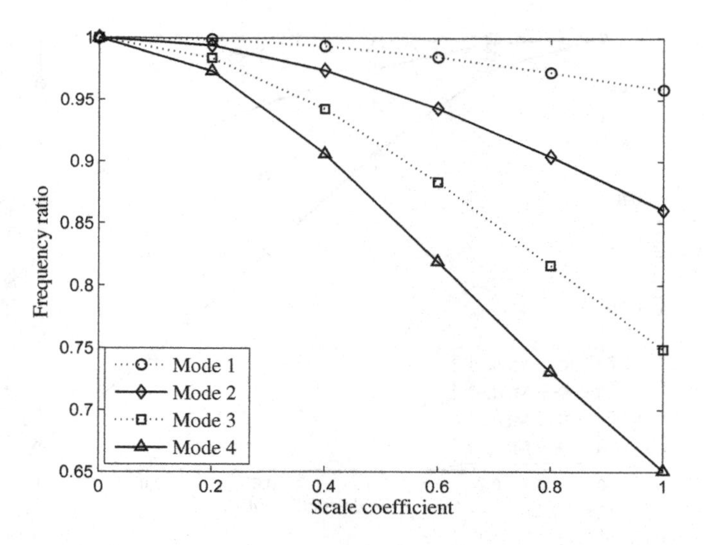

Fig. 6.12 Variation of frequency ratio with e_0a (TBT)

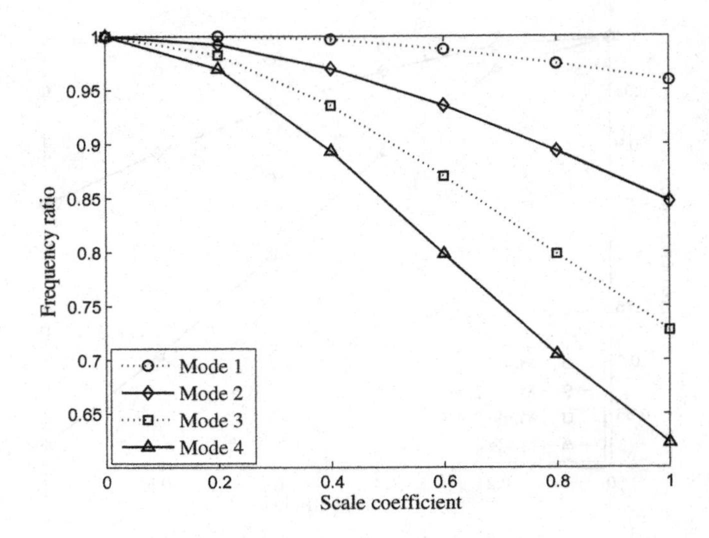

Fig. 6.13 Variation of frequency ratio with e_0a (RBT)

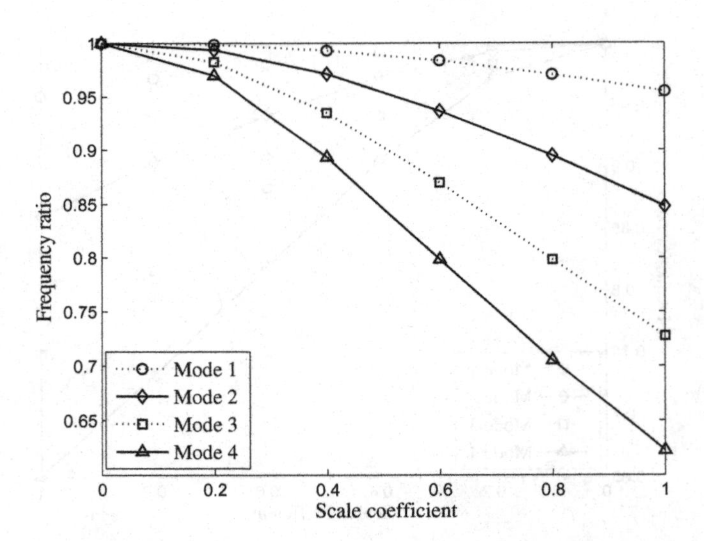

Fig. 6.14 Variation of frequency ratio with e_0a (LBT)

increase in nonlocal parameter in S–S, C–S, C–C boundary conditions, while fundamental frequency parameter increases with nonlocal parameter in the case of cantilever nanobeams. Higher frequency parameters decrease with nonlocal parameter in the case of all boundary conditions. Except fundamental frequency parameter of C–F nanobeam, the frequency parameter associated with nonlocal nanobeams are smaller than the corresponding local nanobeams. This reduction is clearly seen in the case of higher vibration modes. This means that application of local beam models would lead to overprediction of frequency parameters. Hence nonlocal theory should be incorporated for better prediction of higher frequencies of nanobeams.

6.3.4 *Effect of various beam theories*

To investigate the effect of various beam theories such as EBT, TBT, RBT, and LBT on the frequency parameter, variation of fundamental frequency parameter with nonlocal parameter for nanobeams based on various beam theories is shown in Fig. 6.15. In this figure, C–C boundary condition is taken into consideration. It is seen from the figure that EBT predicts higher frequency parameter than other types of beam theories. It is due to the fact that in EBT transverse shear stress and transverse strain are not considered.

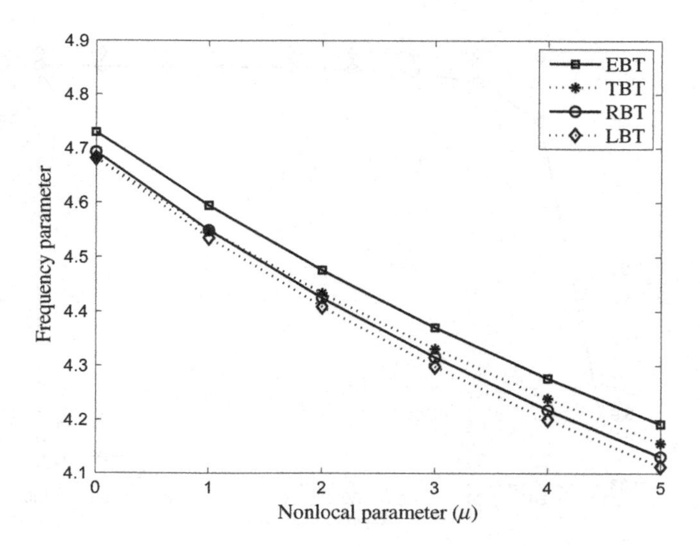

Fig. 6.15 Variation of frequency parameter with nonlocal parameter

Beam theories such as TBT, RBT, and LBT predict approximately closer results. Next we have compared EBT and TBT for understanding the effect of transverse shear deformation and rotary inertia on the vibration frequencies. In contrast to EBT, TBT accounts transverse shear deformation and rotary inertia. To investigate the effects of transverse shear deformation and rotary inertia on the vibration analysis, Fig. 6.16 shows the frequency ratio of nonlocal Timoshenko beam to that of the corresponding nonlocal Euler beam ($\lambda_{NT}/\lambda_{NE}$) with respect to *L/h* for a given scale coefficient of $e_0a = 0.1$ nm. In this figure, we have included the first and fourth modes with S–S boundary condition. It is observed that for all values of *L/h*, frequency ratios are smaller than unity in these modes. This means that frequency parameter obtained by nonlocal TBT is smaller than frequency parameter obtained using nonlocal EBT. This indicates that transverse shear deformation and rotary inertia would lead to reduction of frequencies. One may found that this reduction is seen for higher modes and for small *L/h*. This point is discussed as: Frequency ratio associated with fundamental mode approaches unity in the case of long tubes, while for short tubes (for example, $L/h = 10$), the frequencies of nonlocal Timoshenko and nonlocal Euler beams deviate somewhat from each other. The frequency ratio associated with the fourth mode is significantly smaller than unity especially at

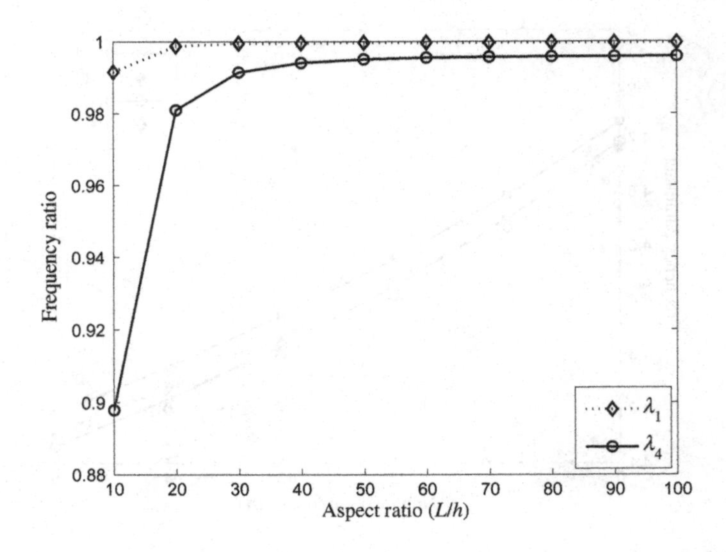

Fig. 6.16 Variation of frequency ratio with *L/h*

small L/h. It is also observed that at higher values of L/h, effects of transverse shear deformation and rotary inertia still have an appreciable effect on the fourth mode. Therefore, the effects of transverse shear deformation and rotary inertia would lead to reduction of frequencies and the reduction is clearly seen at higher modes and also at small L/h. Hence, Timoshenko beam model should be considered when L/h is small and when higher vibration modes are considered.

6.3.5 *Effect of boundary conditions*

In this section, we have considered the effect of boundary condition on the frequency parameter. Figure 6.17 depicts variation of fundamental frequency parameter of TBT nanobeam with nonlocal parameter for different boundary conditions. It is observed from the figure that C–C nanobeams are having the highest frequency parameter and C–F nanobeams are having the lowest frequency parameter. It is also seen that frequency parameters decrease with increase in nonlocal parameter for S–S, C–S, and C–C boundary conditions but frequency parameters increase with nonlocal parameter in the case of C–F nanobeams.

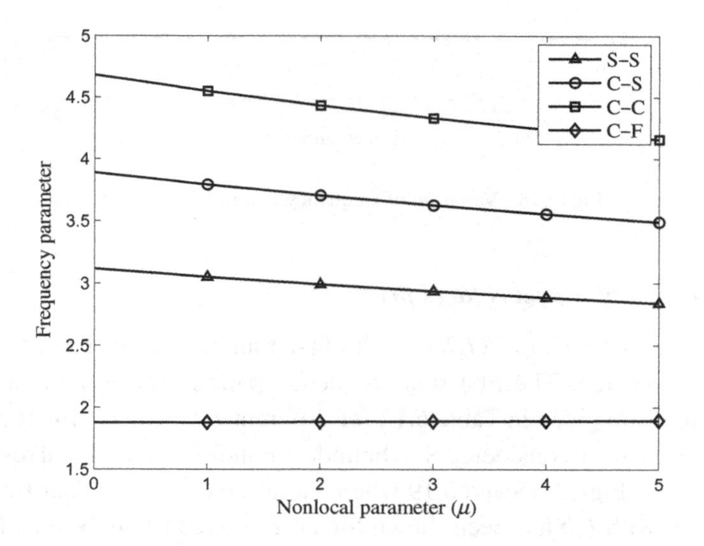

Fig. 6.17 Variation of frequency parameter with nonlocal parameter

Table 6.12 First four frequency parameters of TBT and RBT nanobeams for different *L/h*

	TBT				RBT			
L/h	λ_1	λ_2	λ_3	λ_4	λ_1	λ_2	λ_3	λ_4
10	2.9893	5.3236	6.9697	8.1207	2.9843	5.3015	6.9325	8.0777
12	2.9951	5.3681	7.1001	8.3744	2.9936	5.3642	7.1029	8.3959
14	2.9981	5.3915	7.1713	8.5200	2.9977	5.3925	7.1833	8.5550
16	2.9998	5.4048	7.2129	8.6075	2.9997	5.4071	7.2258	8.6415
18	3.0008	5.4130	7.2386	8.6626	3.0009	5.4153	7.2501	8.6919

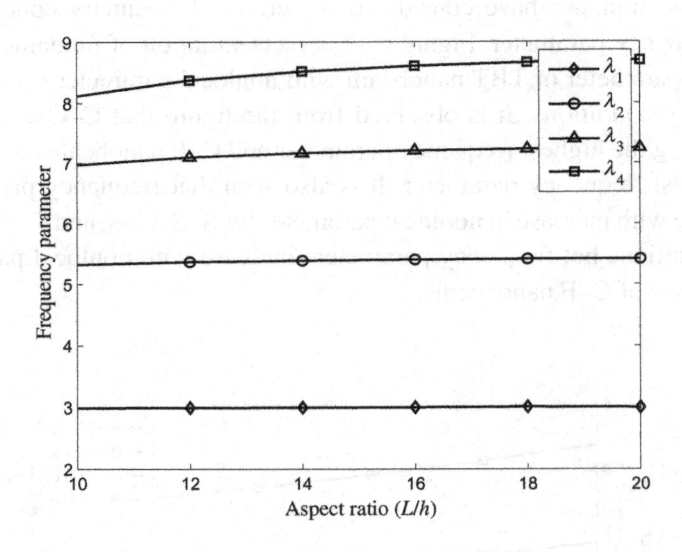

Fig. 6.18 Variation of frequency parameter with *L/h*

6.3.6 *Effect of aspect ratio (L/h)*

Analysis of aspect ratio (*L/h*) on the first four frequency parameters has been investigated. The first four frequency parameters of TBT and RBT nanobeams are given in Table 6.12 for different *L/h* (10, 12, 14, 16, 18). In this table, we have considered S–S boundary condition. Graphical results are illustrated in Figs. 6.18 and 6.19 where variation of the first four frequency parameter with *L/h* has been shown for TBT and RBT nanobeams. In these figures, *L/h* ranges from 10 to 20. It is noticed that frequency parameter

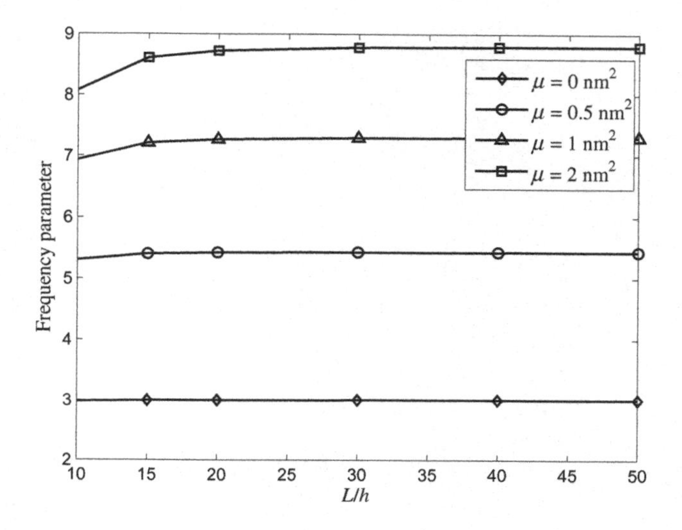

Fig. 6.19 Variation of frequency parameter with *L/h*

increases with increase in *L/h*. It is also seen that the effect of *L/h* is more pronounced for higher vibration modes.

6.4 Conclusions

Vibration characteristics of Euler–Bernoulli and Timoshenko nanobeams have been computed by using simple polynomials and orthonormal polynomials in the Rayleigh–Ritz method. Both tabular and graphical results are given for different scaling effect parameters and boundary conditions. Results are also tabulated for some new boundary conditions (S–F and F–F). Deflection shapes of nonlocal C–C Euler–Bernoulli nanobeams are presented for different scaling effect parameters.

DQM has been applied to investigate free vibration of nanobeams based on different beam theories such as EBT, TBT, RBT, and LBT in conjunction with Eringen's nonlocal elasticity theory. Boundary conditions have been implemented in the coefficient matrix which is quite easy to handle. Effects of nonlocal parameter, boundary condition, aspect ratio, and beam theories on the frequency parameters have been analyzed.

Chapter 7

Vibration of Nanobeams with Complicating Effects

7.1 Vibration of Nanobeams with Non-uniform Material Properties

In this investigation, we have considered carbon nanotube (CNT) with non-uniform material properties which is assumed as per the following relations:

$$E = E_0(1 + pX + qX^2), \quad \rho = \rho_0(1 + rX + sX^2),$$

where E_0 and ρ_0 denote Young's modulus and density at the left end of the CNT, respectively, and p, q, r, and s denote the non-uniform parameters.

Here, we have introduced the following non-dimensional parameters:

$$X = \frac{x}{L},$$

$$W = \frac{w_0}{L},$$

$$\alpha = \frac{e_0 a}{L} = \text{scaling effect parameter},$$

$$\xi = \frac{L\sqrt{A}}{I},$$

$$\tau = \frac{1}{\xi^2},$$

$$\lambda^2 = \frac{\rho_0 A\omega^2 L^4}{E_0 I} = \text{frequency parameter},$$

$$\Omega = \frac{EI_0}{k_s GAL^2} = \text{shear deformation parameter}.$$

To apply Raleigh–Ritz method, one needs Rayleigh quotient which is obtained by equating maximum kinetic and strain energies. As such, one may obtain Rayleigh quotient in non-dimensional form for Euler–Bernoulli beam theory (EBT) as

$$\lambda^2 = \frac{\int_0^1 (1 + pX + qX^2)\left(\frac{d^2W}{dX^2}\right)^2 dX}{\int_0^1 (1 + rX + sX^2)\left(W^2 - \alpha^2 W \frac{d^2W}{dX^2}\right) dX}. \tag{7.1}$$

Similarly, Rayleigh quotient in non-dimensional form for Timoshenko beam theory (TBT) is obtained as

$$\lambda^2 = \frac{\int_0^1 (1 + pX + qX^2)\left(\left(\frac{d\phi}{dX}\right)^2 + \frac{1}{\Omega}\left(\phi + \frac{dW}{dX}\right)^2\right) dX}{\int_0^1 (1 + rX + sX^2)\left(W^2 + \tau\phi^2 + \alpha^2 W \frac{d\phi}{dX} + \tau\alpha^2 \left(\frac{d\phi}{dX}\right)^2\right) dX}. \tag{7.2}$$

As such, matrices K and M_a for EBT are given as below:

$$K(i,j) = \int_0^1 (1 + pX + qX^2)\varphi_i''\varphi_j''\, dX,$$

$$M_a(i,j) = \int_0^1 (1 + rX + sX^2)\varphi_i\varphi_j - \frac{\alpha^2}{2}\varphi_i\varphi_j'' - \frac{\alpha^2}{2}\varphi_i''\varphi_j\, dX.$$

Similarly, matrices K and M_a for TBT are given as

$$K = \begin{bmatrix} k_1 & k_2 \\ k_3 & k_4 \end{bmatrix}.$$

Here k_1, k_2, k_3, and k_4 are submatrices and are given by

$$k_1(i,j) = \int_0^1 (1 + pX + qX^2)\varphi_i'\varphi_j\, dX,$$

$$k_2(i,j) = \int_0^1 (1 + pX + qX^2)\varphi_i'\psi_j\, dX,$$

$$k_3(i,j) = \int_0^1 (1 + pX + qX^2)\psi_i \varphi_j' \, dX,$$

$$k_4(i,j) = \int_0^1 (1 + pX + qX^2)(\psi_i \psi_j + \Omega \psi_i' \psi_j') \, dX,$$

$$M_a = \begin{bmatrix} m_1 & m_2 \\ m_3 & m_4 \end{bmatrix}.$$

Submatrices m_1, m_2, m_3, and m_4 are as follows:

$$m_1(i,j) = \Omega \int_0^1 (1 + rX + sX^2)\varphi_i \varphi_j \, dX,$$

$$m_2(i,j) = \Omega \frac{\alpha^2}{2} \int_0^1 (1 + rX + sX^2)\varphi_i \psi_j' \, dX,$$

$$m_3(i,j) = \Omega \frac{\alpha^2}{2} \int_0^1 (1 + rX + sX^2)\psi_i' \varphi_j \, dX,$$

$$m_4(i,j) = \Omega \int_0^1 (1 + rX + sX^2)(\tau \psi_i \psi_j + \tau \alpha^2 \psi_i' \psi_j') \, dX.$$

7.2 Numerical Results and Discussions

In the numerical evaluations, following parameters of single-walled carbon nanotubes (SWCNTs) have been used (Wang *et al.* 2007): diameter, $d = 0.678$ nm; length, $L = 10d$; thickness, $t = 0.066$; shear correction factor, $k_s = 0.563$; Young's modulus, $E_0 = 5.5$ TPa; shear modulus, $G = E_0/[2(1+\nu)]$; Poisson's ratio $\nu = 0.19$; and second moment of area $I = \Pi d^4/64$.

7.2.1 *Convergence of the method*

A convergence study has been shown in Tables 7.1 and 7.2, respectively, for EBT and TBT. In these tables, we have shown convergence of first three frequency parameters ($\sqrt{\lambda}$). Here, we have taken non-uniform parameters as $p = q = r = s = 0.1$ and scaling effect parameter as 0.3. Convergency has been reported for S–S and C–S edge conditions. It is clearly seen from the table that convergency is achieved as we increase the number of terms. One may notice that $n = 11$ is sufficient for computing the results.

Table 7.1 Convergence of the first three frequency parameters of EBT nanobeams

	S–S			C–S		
n	λ_1	λ_2	λ_3	λ_1	λ_2	λ_3
3	2.6803	4.7956	6.4334	3.2675	4.8388	7.2365
4	2.6801	4.3104	6.4307	3.2643	4.7903	6.0470
5	2.6797	4.3101	5.4811	3.2643	4.7640	5.9176
6	2.6797	4.3018	5.4807	3.2643	4.7630	5.8429
7	2.6797	4.3018	5.4432	3.2643	4.7627	5.8367
8	2.6797	4.3018	5.4432	3.2643	4.7627	5.8344
9	2.6797	4.3018	5.4426	3.2643	4.7627	5.8343
10	2.6797	4.3018	5.4426	3.2643	4.7627	5.8343
11	2.6797	4.3018	5.4426	3.2643	4.7627	5.8343

Table 7.2 Convergence of the first three frequency parameters of TBT nanobeams

	S–S			C–S		
n	λ_1	λ_2	λ_3	λ_1	λ_2	λ_3
3	2.7683	4.5273	10.3126	3.0810	5.0918	7.7717
4	2.6394	4.5264	5.7314	3.1743	4.5316	6.1565
5	2.6394	4.1354	5.7298	3.1720	4.4990	5.4307
6	2.6390	4.1354	5.0898	3.1720	4.4814	5.3652
7	2.6390	4.1286	5.0898	3.1720	4.4807	5.3184
8	2.6390	4.1286	5.0627	3.1720	4.4805	5.3151
9	2.6390	4.1286	5.0627	3.1720	4.4805	5.3139
10	2.6390	4.1286	5.0622	3.1720	4.4805	5.3138
11	2.6390	4.1286	5.0622	3.1720	4.4805	5.3138

7.2.2 *Validation*

For the validation purpose, we have considered an uniform ($p = q = r = s = 0$) nanobeam. To compare our results with that of Wang *et al.* (2007), we have taken same parameters as that of Wang *et al.* (2007). Table 7.3 shows comparison of the first three frequency parameters of Euler–Bernoulli and Timoshenko nanobeams for simply-supported boundary conditions. Results have been compared with scaling effect parameters as 0, 0.3, and 0.5. It is noticed from the table that frequency parameters ($\sqrt{\lambda}$) decrease with

Table 7.3 Comparison of frequency parameters for uniform nanobeams

$\alpha = 0$		$\alpha = 0.3$		$\alpha = 0.5$	
Present	Ref.*	Present	Ref.*	Present	Ref.*
EBT					
3.1416	3.1416	2.6800	2.6800	2.3022	2.3022
6.2832	6.2832	4.3013	4.3013	3.4604	3.4604
9.4248	9.4248	5.4422	5.4422	4.2941	4.2941
15.5665	15.708	6.3630	6.3630	4.9820	4.9820
TBT					
3.0742	3.0929	3.0072	3.0243	2.6412	2.6538
5.8274	5.9399	5.4400	5.5304	4.1357	4.2058
8.1757	8.4444	7.2662	7.4699	5.0744	5.2444

*Wang *et al.* (2007).

increase in scaling effect parameter. From this table, one may observe close agreement of the results with that of available literature.

7.2.3 *Effect of non-uniform parameter*

In this section, we have studied the effects of non-uniform parameters on the frequency parameters. Four cases are considered here which are discussed below.

Case 1: Density varies with space coordinate and Young's modulus as constant, viz. $\rho = \rho_0(1 + rX + sX^2)$ and $E = E_0$. Here following subcases may arise:

(a) $s = 0$ and $r \neq 0$.
(b) $r = 0$ and $s \neq 0$.
(c) $r \neq 0$ and $s \neq 0$.

Case 2: Young's modulus varies with space coordinate and density as constant, viz. $E = E_0(1 + pX + qX^2)$ and $\rho = \rho_0$. Here also following three

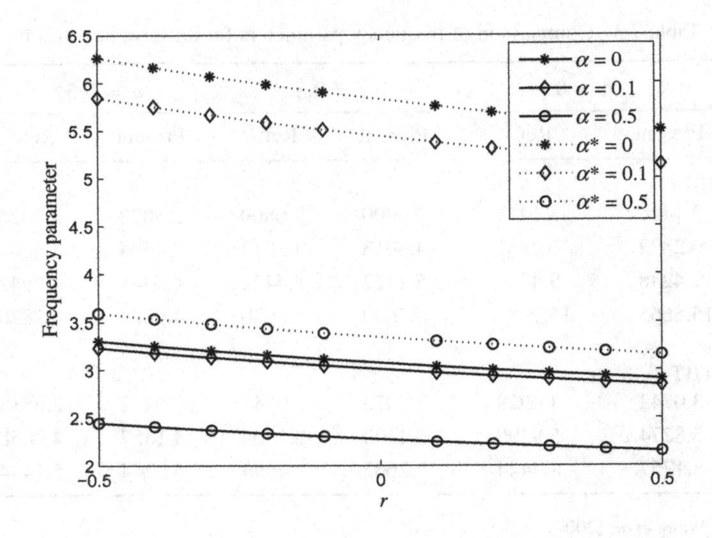

Fig. 7.1 Variation of frequency parameters with r

subcases may arise:

(a) $p = 0$ and $q \neq 0$.
(b) $q = 0$ and $p \neq 0$.
(c) $p \neq 0$ and $q \neq 0$.

Case 3: Young's modulus and density both vary with space coordinate, viz. $E = E_0(1 + pX + qX^2)$ and $\rho = \rho_0(1 + rX + sX^2)$. Four subcases arise here are:

(a) Linear variations of Young's modulus and density.
(b) Quadratic variations of Young's modulus and density.
(c) Linear variation of Young's modulus and quadratic variation of density.
(d) Linear variation of density and quadratic variation of Young's modulus.

Figures 7.1 and 7.2 are the pictorial representation of Cases 1(a) and 1(b), respectively. Here, we have considered TBT nanobeams to show the behavior of the first two frequency parameters with r and s, respectively, for the scaling effect parameters as 0, 0.1, and 0.5. In Fig. 7.1, we have considered S–S boundary condition and in Fig. 7.2, we have taken C–S boundary condition. The solid and dotted lines represent the first and second frequency parameters, respectively. Also α and α^* depict scaling effect parameters for

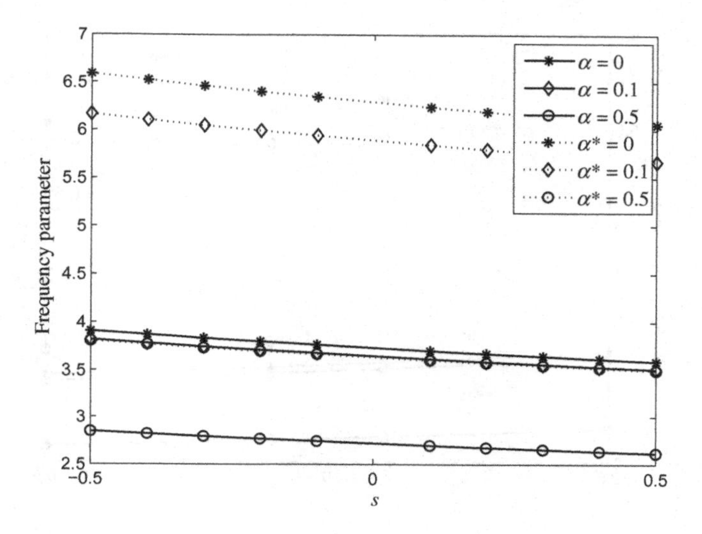

Fig. 7.2 Variation of frequency parameters with s

the first and second frequency parameters, respectively. Varying parameter varies from -0.5 to 0.5. In these figures, it can be clearly seen that frequency parameters decrease with r and s.

To show the results of Case 1(c), we have given the results in two ways. Firstly, keeping r constant, varying s and secondly, keeping s constant, varying r. For these two cases, the results are illustrated in Figs. 7.3 and 7.4, respectively, for S–S boundary condition. From these figures also, we can conclude that if we fix one and other vary, then the frequency parameters decrease with the increase of varying parameter.

Figures 7.5 and 7.6 are the graphical depiction of Cases 2(a) and 2(b), respectively. Graphical results have been shown for TBT nanobeams with S–S boundary condition. Varying parameter varies from -0.5 to 0.5. Graphs are drawn for the first two frequency parameters. It is observed that the frequency parameters increase with q and p. The results of Case 2(c) are given in Fig. 7.7 where p is fixed, q varies and in Fig. 7.8 where q is fixed, p varies. In these graphs, we have considered C–S boundary condition. In these graphs, it is observed that the frequency parameters increase with increase of varying parameter.

Results of Case 3 are given for EBT nanobeams and four subcases are considered. First, we have shown the effects of non-uniform parameters

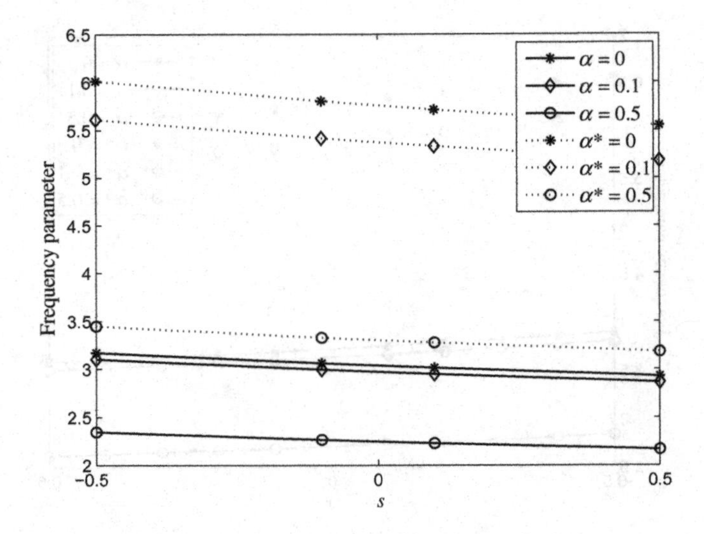

Fig. 7.3 Variation of frequency parameters with s

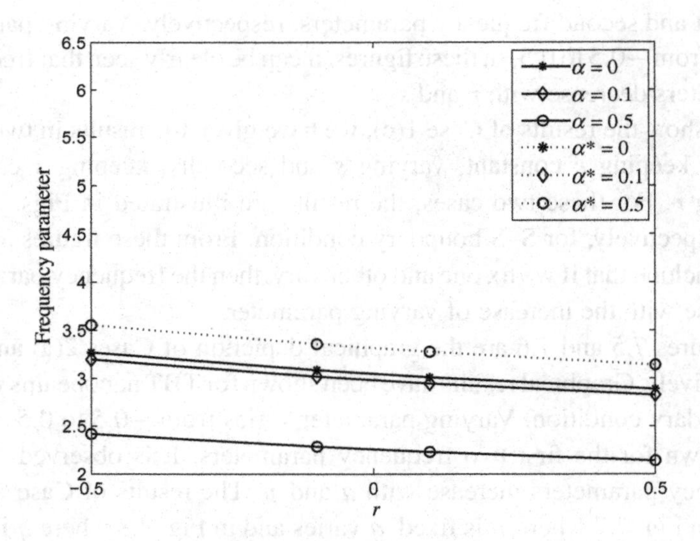

Fig. 7.4 Variation of frequency parameters with r

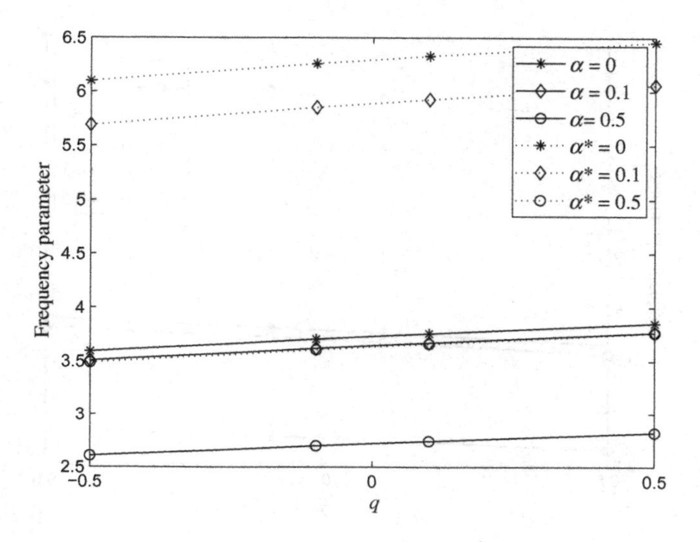

Fig. 7.5 Variation of frequency parameters with q

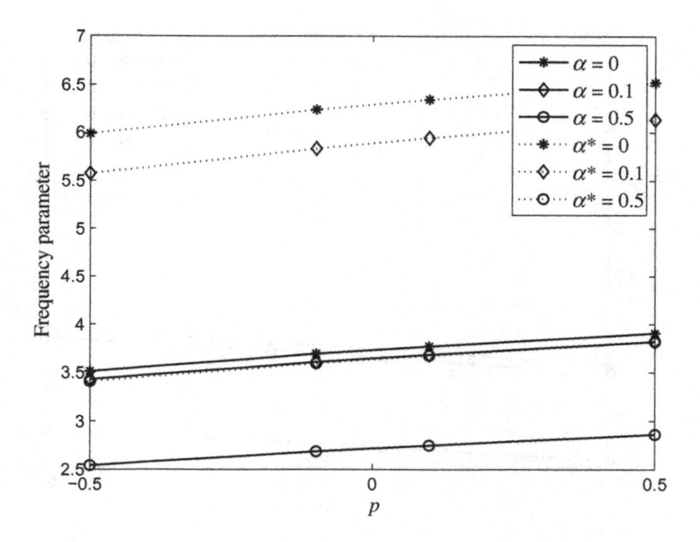

Fig. 7.6 Variation of frequency parameters with p

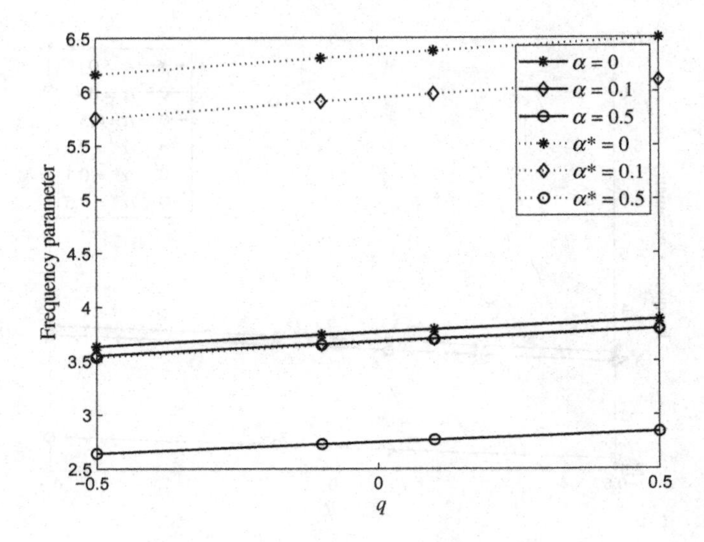

Fig. 7.7 Variation of frequency parameters with q

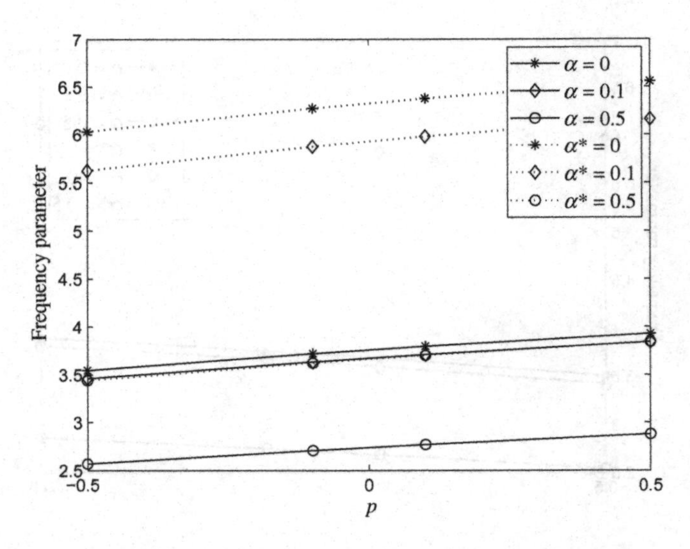

Fig. 7.8 Variation of frequency parameters with p

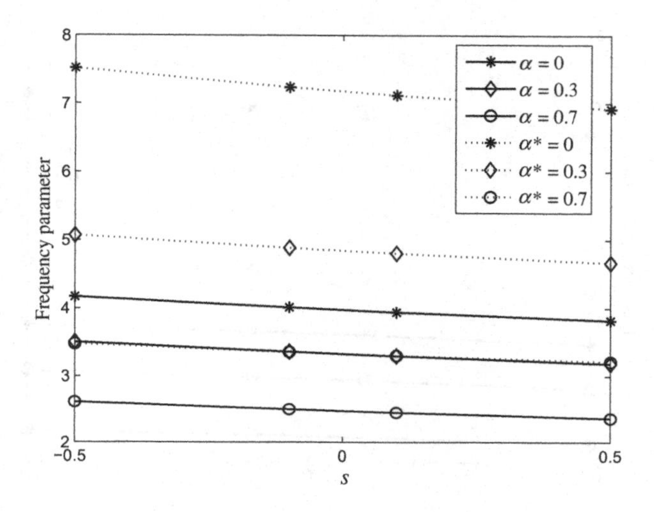

Fig. 7.9 Variation of frequency parameters with s

when density and Young's modulus vary quadratically. This case may be achieved by assigning zero to p and r. Variation of first two frequency parameters with s has been illustrated in Fig. 7.9 keeping q constant (0.1). Similarly, effect of q on the first two frequency parameters has been shown in Fig. 7.10 keeping s constant (0.1). In these graphs, varying parameters range from −0.5 to 0.5 with C–S edge condition. Results have been shown for different values of scaling effect parameters (0, 0.3, and 0.7). One may see that frequency parameters increase with q and decrease with s. It is also observed that frequency parameters decrease with increase in scaling effect parameter. This means that frequency parameters are overpredicted when we consider local beam model for vibration analysis of nanobeams. It is also observed that frequency parameters increase with increase in mode number.

In this paragraph, we have presented the effects non-uniform parameters when density and Young's modulus vary linearly. This is achieved by taking q and s to zero. Graphical variation of frequency parameters with p taking r constant (0.2) has been shown in Fig. 7.11. Similarly, graphical variation of frequency parameters with r taking p constant (0.2) has been shown in Fig. 7.12. The graphs are plotted for different values of scaling effect parameters with C–S boundary condition. It is noticed that frequency parameters increase with p and decrease with r.

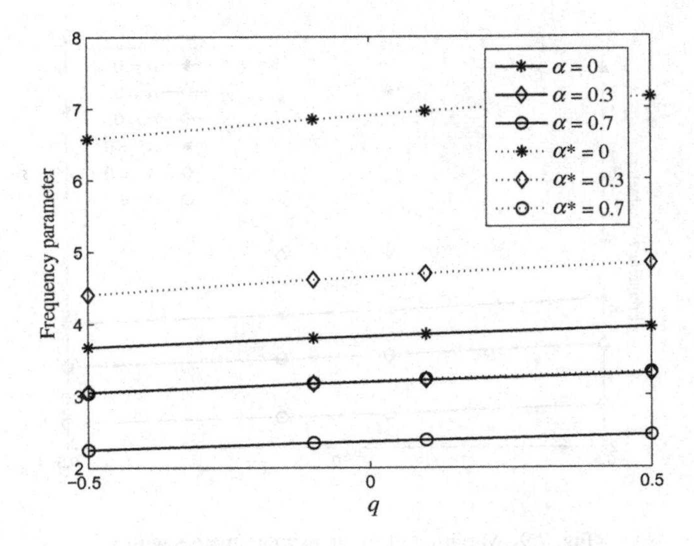

Fig. 7.10 Variation of frequency parameters with q

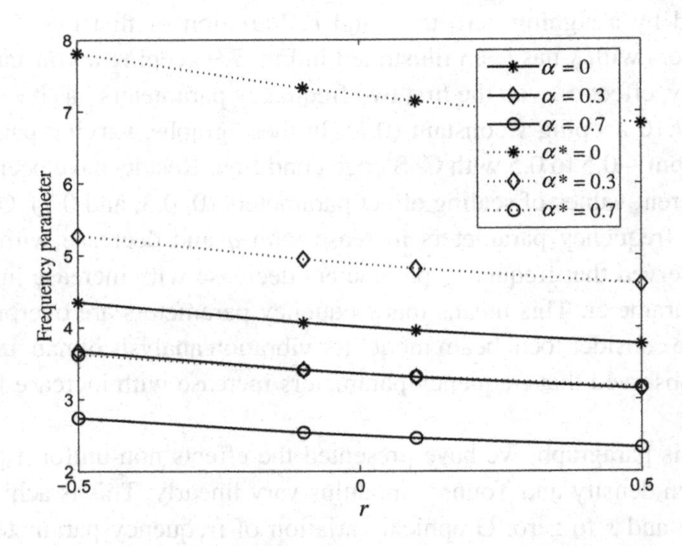

Fig. 7.11 Variation of frequency parameters with r

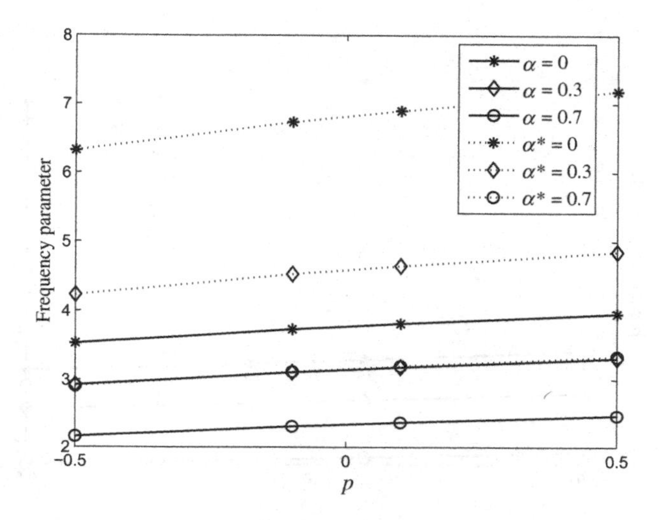

Fig. 7.12 Variation of frequency parameters with p

Here, we have considered the effects of non-uniform parameters when Young's modulus varies linearly and density varies quadratically. This is the situation which is obtained by taking q and r as zero. Figures 7.13 and 7.14 depict variation of frequency parameters with s and p, respectively. In these graphs, we have taken S–S boundary condition with non-varying parameter as 0.1. Results have been shown for different values of scaling effect parameters. In these graphs, one may observe that frequency parameters decrease with s and increase with p.

Next, we have analyzed the effects of non-uniform parameter when Young's modulus varies quadratically and density varies linearly. For this, we have taken p and s as zero. Variations of the first two frequency parameters with q and r have been given in Figs. 7.15 and 7.16. Results have been shown with the non-varying parameter as 0.1 and boundary condition as S–S. In these graphs, one may see that frequency parameters decrease with r and increase with q.

7.2.4 *Effect of small scale parameter*

To investigate the behavior of scaling effect parameter on the frequency parameters, variation of frequency ratio for EBT nanobeams with scale

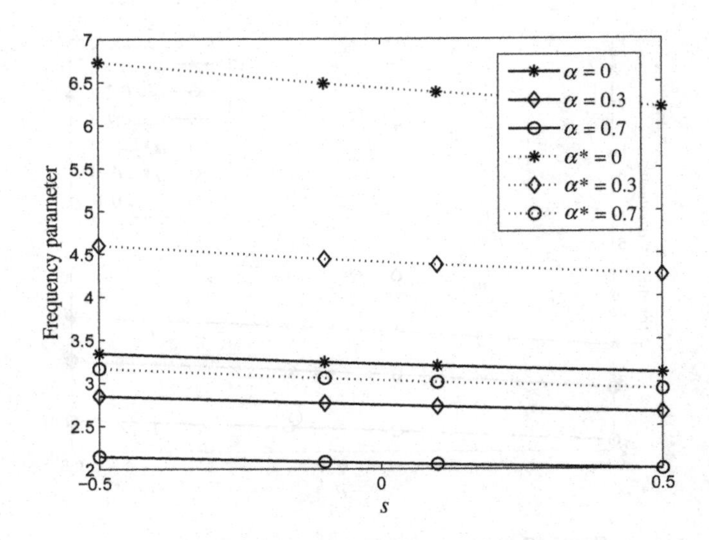

Fig. 7.13 Variation of frequency parameters with s

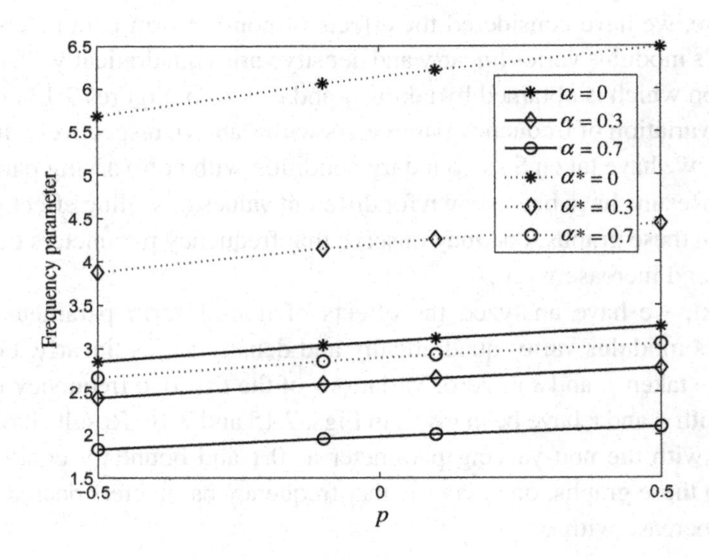

Fig. 7.14 Variation of frequency parameters with p

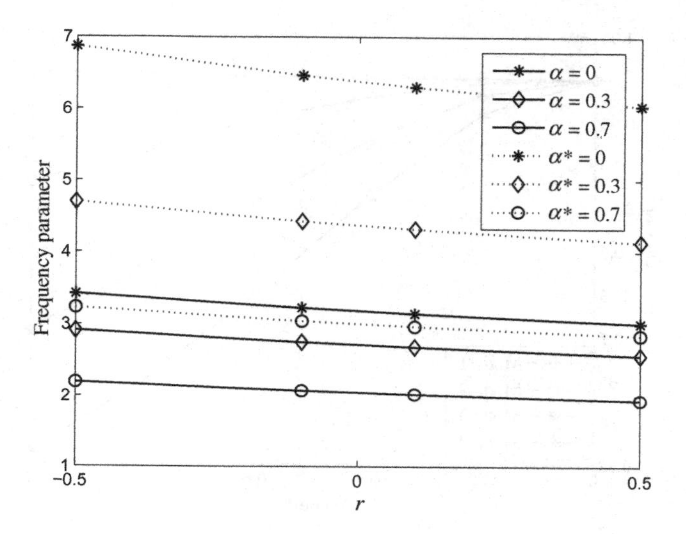

Fig. 7.15 Variation of frequency parameters with r

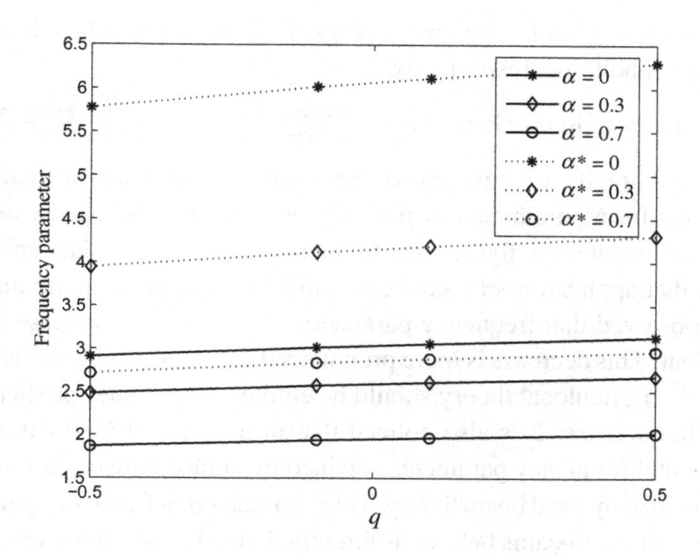

Fig. 7.16 Variation of frequency parameters with q

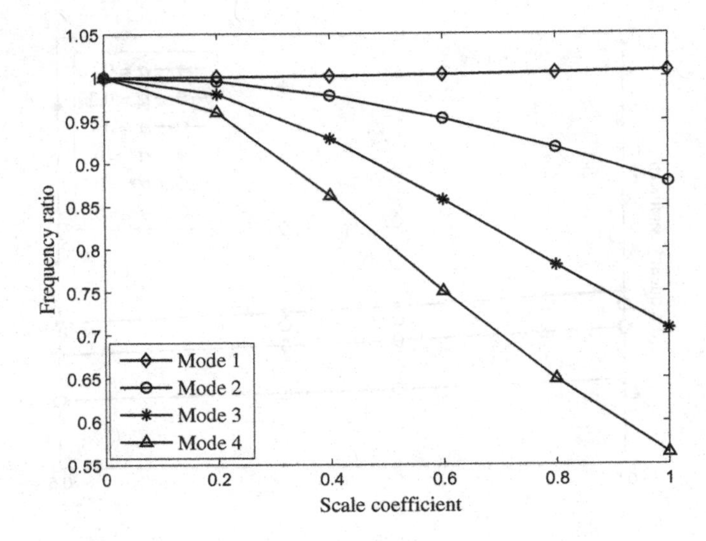

Fig. 7.17 Variation of frequency ratio with e_0a (C–F)

coefficient (e_0a) has been shown in Figs. 7.17 and 7.18 with C–F and S–F boundary conditions, respectively.

Frequency ratio is calculated as $\frac{\text{frequency parameter calculated using nonlocal theory}}{\text{frequency parameter calculated using local theory}}$.

In these graphs, we have shown the results for the first four modes with the non-uniform parameters as $p = 0.1, q = 0.2, r = 0.3$, and $s = 0.4$. It is observed from these figures that frequency ratios are less than unity. This implies that application of local beam model overpredicts the frequency. It is also observed that frequency parameters decrease with increase in scale coefficient. This decrease is more pronounced in the case of higher vibration modes. Thus, nonlocal theory should be employed for better predictions of higher frequencies. It is also noticed that in the case of C–F nanobeams, fundamental frequency parameter obtained by nonlocal theory is more than that furnished by local beam theory. Thus, fundamental frequency parameter of cantilever nanobeams behave differently than other boundary conditions.

7.2.5 *Effect of boundary condition*

Effect of boundary condition on the frequency parameters is investigated. Variation of fundamental frequency parameter with α is shown in Fig. 7.19

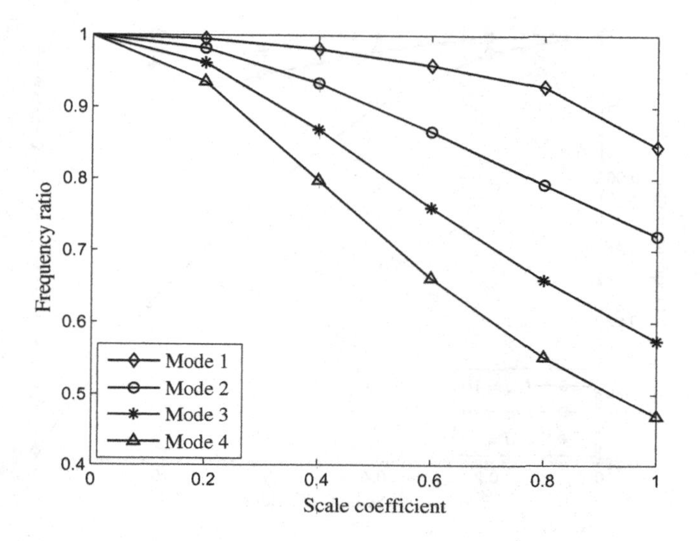

Fig. 7.18 Variation of frequency ratio with e_0a (S–F)

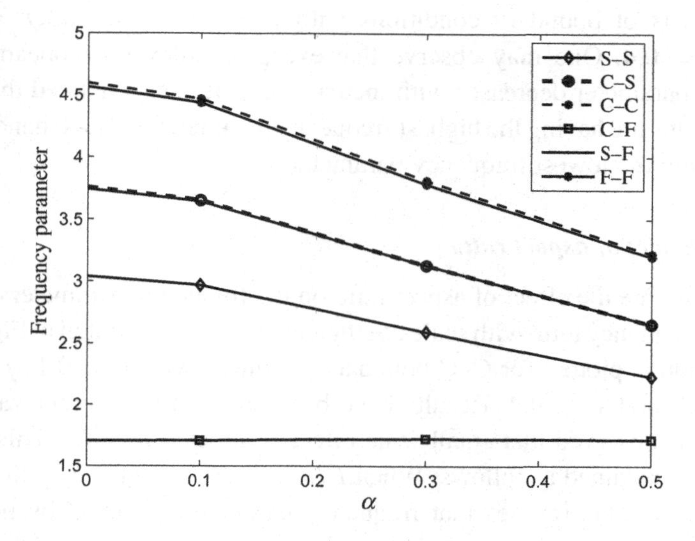

Fig. 7.19 Variation of frequency parameter with α

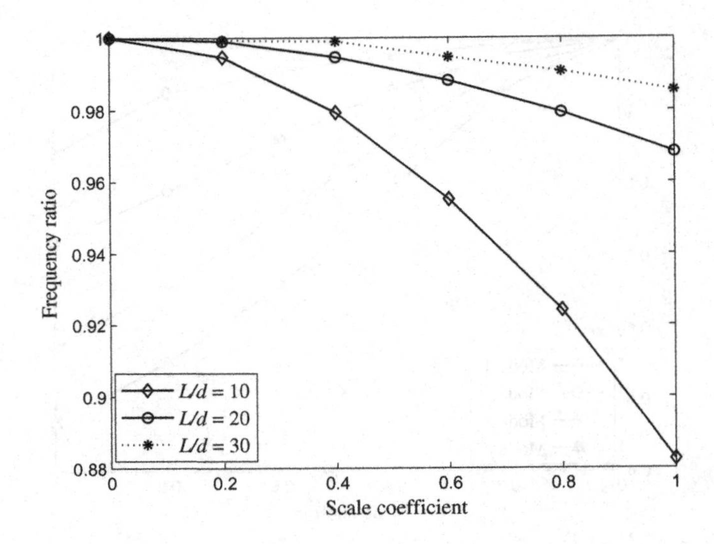

Fig. 7.20 Variation of frequency ratio with $e_o l_{\text{int}}$

for all sets of boundary conditions with $p = 0.1, q = 0.2, r = 0.3$, and $s = 0.4$. One may observe that except cantilever nanobeams, frequency parameter decreases with increase in α. It is also noticed that C–C nanobeams are having the highest frequency parameter and C–F nanobeams are having the lowest frequency parameter.

7.2.6 *Effect of aspect ratio*

To investigate the effect of aspect ratio on the frequency parameters, variation of frequency ratio with scale coefficient has been illustrated in Fig. 7.20. The graph is plotted for C–C boundary condition with $p = 0.1, q = 0.2$, $r = 0.3$, and $s = 0.4$. Results have been shown for different values of L/d. It is observed that small scale effect is affected by L/d. This observation is explained as follows: When L/d increases, frequency ratio comes closer to 1. This implies that frequency parameter obtained by nonlocal beam model comes closer to that furnished by local beam model. Hence, small scale effect is negligible for a very slender CNT, while it is significant for short CNTs. This implies that if we compare magnitude of small scale effect with length of the slender tube, the small scale (internal characteristic length) is so small that it can be regarded as zero. Next, we have shown

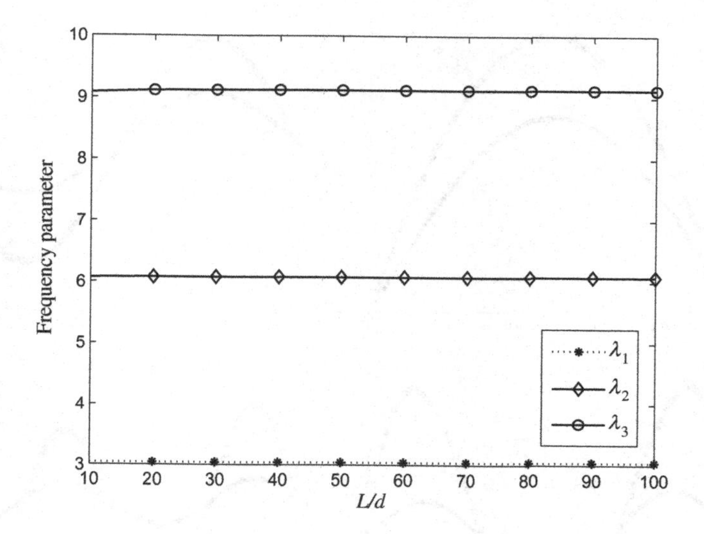

Fig. 7.21 Variation of frequency parameter with L/d

variation of the first three frequency parameters with L/d in Fig. 7.21 with $p = 0.1, q = 0.2, r = 0.3$, and $s = 0.4$. In this figure, μ is taken as 0.1 nm^2 and L/d ranges from 10 to 100, where d is assumed to be 0.678 nm. In this graph, one may notice that frequency parameter increases with L/d.

7.2.7 *Mode shapes*

First few higher mode shapes are given in Figs. 7.22 and 7.23 for comparing the results which may help the researchers. Here the first four mode shapes of S–S and C–S boundary conditions are given with scaling effect parameters as 0, 0.3, and non-uniform parameters as $p = 0.1, q = 0.2, r = 0.3$, and $s = 0.4$. One may observe that mode shapes are affected by the scaling effect parameter.

7.2.8 *Vibration analysis of nanobeams embedded in elastic foundations*

Here, we have investigated vibration of embedded nanobeams in thermal environments based on EBT, TBT, and Reddy–Bickford beam theory (RBT) beam theories. The nanobeam is embedded in elastic foundations such as Winkler and Pasternak. Rayleigh–Ritz has been applied in EBT and TBT

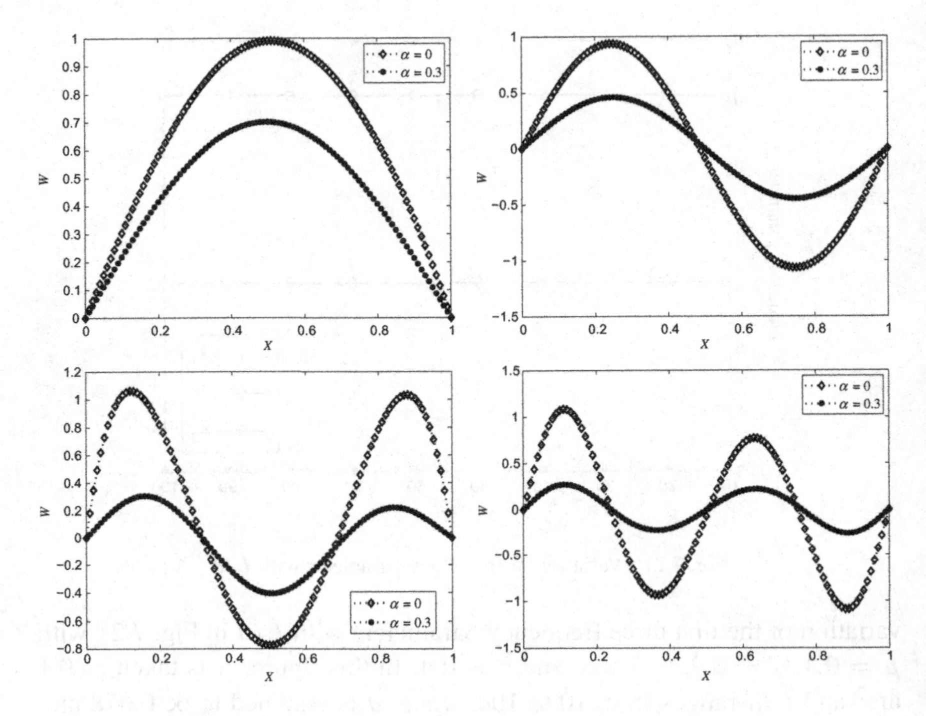

Fig. 7.22 First four deflection shapes of S–S Euler–Bernoulli nanobeams

with shape functions as boundary characteristic orthogonal polynomials and Chebyshev polynomials, respectively. Differential quadrature method (DQM) has been employed in vibration of embedded nanobeams based on RBT.

For simplicity and convenience in mathematical formulation, the following non-dimensional parameters are introduced:

$$X = \frac{x}{L}, \quad W = \frac{w_0}{L}, \quad \alpha = \frac{e_0 a}{L}, \quad \tau = \frac{I}{AL^2}, \quad \Omega = \frac{EI}{k_s GAL^2},$$

$$\bar{\Omega} = \frac{G\tilde{A}L^2}{EI}, \quad \lambda^2 = \frac{\rho A \omega^2 L^4}{EI}, \quad K_g = \frac{k_g L^2}{EI},$$

$$K_w = \frac{k_w L^4}{EI}, \quad \hat{N}_\theta = \frac{N_\theta L^2}{EI}.$$

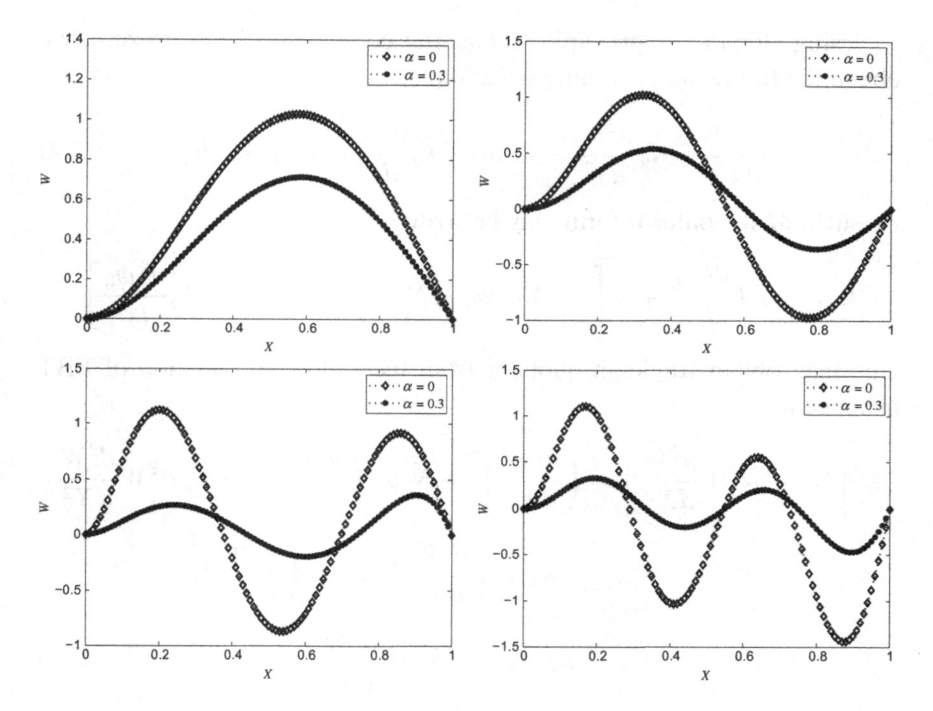

Fig. 7.23 First four deflection shapes of C–S Euler–Bernoulli nanobeams

7.3 Euler–Bernoulli Beam Theory (EBT)

Maximum strain energy $U_{\max}$ may be given as Eq. (3.18). Maximum kinetic energy is written as Eq. (3.20).

Maximum potential energy due to the axial force may be expressed as

$$V_a = \frac{1}{2}\int_0^L \left\{ \bar{N}\left(\frac{dw_0}{dx}\right)^2 + f_e w_0 \right\} dx, \tag{7.3}$$

where the parameters have already been defined in Section 3.2.7.1. Here again, we have defined $\bar{N}$ as the axial force which is expressed as $\bar{N} = N_m + N_\theta$. It is noted that N_m will be taken as zero for vibration analysis.

Using Hamilton's principle and setting coefficient of δw_0 to zero, we obtain the following governing equation:

$$\frac{d^2M}{dx^2} + N_\theta \frac{d^2 w_0}{dx^2} - k_w w_0 + k_g \frac{d^2 w_0}{dx^2} = -\rho A \omega^2 w_0. \tag{7.4}$$

As such, M in nonlocal form may be written as

$$M = -EI\frac{d^2 w_0}{dx^2} + \mu \left[-\rho A \omega^2 w_0 - N_\theta \frac{d^2 w_0}{dx^2} + k_w w_0 - k_g \frac{d^2 w_0}{dx^2} \right].$$

One may obtain Rayleigh quotient from the following equation of EBT nanobeams:

$$\begin{aligned}\lambda^2 \left[W^2 - \alpha^2 W \frac{d^2 W}{dX^2} \right] = \left(\frac{d^2 W}{dX^2} \right)^2 + \hat{N}_\theta \alpha^2 \left(\frac{d^2 W}{dX^2} \right)^2 - K_w \alpha^2 W \frac{d^2 W}{dX^2} \\ + K_g \alpha^2 \left(\frac{d^2 W}{dX^2} \right)^2 + \hat{N}_\theta \left(\frac{d^2 W}{dX^2} \right)^2 \\ + K_w W^2 - K_g W \frac{d^2 W}{dX^2}.\end{aligned} \tag{7.5}$$

Here, we have used orthonormal polynomials ($\hat{\varphi_k}$) in Eq. (3.5). Substituting Eq. (3.5) in Eq. (7.5) and minimizing λ^2 with respect to constant coefficients, the following eigenvalue value problem may be obtained:

$$[K]\{Z\} = \lambda^2 [M_a]\{Z\}, \tag{7.6}$$

where Z is a column vector of constants and stiffness matrix K as well as mass matrix M_a are given by

$$\begin{aligned}K(i,j) = \textstyle\int_0^1 ((2 + 2K_g\alpha^2 + 2\hat{N}_\theta\alpha^2)\phi_i''\phi_j'' - K_w\alpha^2\phi_i''\phi_j - K_w\alpha^2\phi_i\phi_j'' \\ + 2K_w\phi_i\phi_j - K_g\phi_i''\phi_j - K_g\phi_i\phi_j'' + 2\hat{N}_\theta\alpha^2\phi_i'\phi_j')dX,\end{aligned}$$

$$M_a(i,j) = \textstyle\int_0^1 (2\phi_i\phi_j - \alpha^2\phi_i''\phi_j - \alpha^2\phi_i\phi_j'')dX.$$

7.4 Timoshenko Beam Theory (TBT)

Maximum strain energy $U_{\max}$ may be given as Eq. (3.22). Maximum kinetic energy is written as Eq. (3.25). Maximum potential energy due to work done may be written as Eq. (7.3).

Applying Hamilton's principle and setting coefficient of δw_0 and $\delta\phi_0$ to zero, governing equilibrium equations are obtained as

$$\frac{dM}{dx} - Q = -\rho I \omega^2 \phi_0, \tag{7.7}$$

$$\frac{dQ}{dx} + N_\theta \frac{d^2 w_0}{dx^2} - f_e = -\rho A \omega^2 w_0. \tag{7.8}$$

Using Eqs. (7.7)–(7.11) and Eqs. (1.13) and (1.14), one may obtain bending moment M and shear force Q in nonlocal form as follows:

$$M = EI\frac{d\phi_0}{dx} + \mu\left[-\rho I \omega^2 \frac{d\phi_0}{dx} - \rho A\omega^2 w_0 - N_\theta \frac{d^2 w_0}{dx^2} + f_e\right], \tag{7.9}$$

$$\begin{aligned} Q &= k_s GA\left(\phi_0 + \frac{dw_0}{dx}\right) \\ &\quad + \mu\left[-\rho A \omega^2 \frac{dw_0}{dx} - N_\theta \frac{d^3 w_0}{dx^3} + k_w \frac{dw_0}{dx} - k_g \frac{d^3 w_0}{dx^3}\right]. \end{aligned} \tag{7.10}$$

Equating maximum kinetic and potential energies, one may obtain the following expressions for TBT nanobeams:

$$\begin{aligned} &\lambda^2\left[W^2 + \tau\phi^2 + \tau\alpha^2\left(\frac{d\phi}{dX}\right)^2 + \alpha^2 W\frac{d\phi}{dX} + \alpha^2\frac{dW}{dX}\left(\phi + \frac{dW}{dX}\right)\right] \\ &= \left(\frac{d\phi}{dX}\right)^2 - \hat{N}_\theta \alpha^2 \frac{d\phi}{dX}\frac{d^2 W}{dX^2} - K_w \alpha^2 W \frac{d\phi}{dX} \\ &\quad - K_g \alpha^2 \frac{d^2 W}{dX^2}\frac{d\phi}{dX} + \frac{1}{\Omega}\left(\phi + \frac{dW}{dX}\right)^2 \\ &\quad - \hat{N}_\theta \alpha^2 \frac{d^3 W}{dX^3}\left(\phi + \frac{dW}{dX}\right) + K_w \alpha^2 \frac{dW}{dX}\left(\phi + \frac{dW}{dX}\right) \\ &\quad - K_g \alpha^2 \frac{d^3 W}{dX^3}\left(\phi + \frac{dW}{dX}\right) + \hat{N}_\theta\left(\frac{dW}{dX}\right)^2 \\ &\quad + K_w W^2 - K_g W \frac{d^2 W}{dX^2}. \end{aligned} \tag{7.11}$$

In this problem, we have used Chebyshev polynomials in the Rayleigh–Ritz method. As such, we introduce another independent variable ξ as

$\xi = 2X - 1$ which transforms the range $0 \leq X \leq 1$ into the applicability range $-1 \leq \xi \leq 1$.

Now, Rayleigh quotient may be obtained from the following equations of TBT nanobeams:

$$\lambda^2 \left[W^2 + \tau\phi^2 + 4\tau\alpha^2 \left(\frac{d\phi}{d\xi}\right)^2 + 2\alpha^2 W \frac{d\phi}{d\xi} + 2\alpha^2 \frac{dW}{d\xi}\left(\phi + 2\frac{dW}{d\xi}\right)\right]$$
$$= 4\left(\frac{d\phi}{d\xi}\right)^2 - 8\hat{N}_\theta \alpha^2 \frac{d\phi}{d\xi}\frac{d^2W}{d\xi^2} + 2K_w\alpha^2 W \frac{d\phi}{d\xi}$$
$$- 8K_g\alpha^2 \frac{d^2W}{d\xi^2}\frac{d\phi}{d\xi} + \frac{1}{\Omega}\left(\phi + 2\frac{dW}{d\xi}\right)^2$$
$$- 8\hat{N}_\theta \alpha^2 \frac{d^3W}{d\xi^3}\left(\phi + 2\frac{dW}{d\xi}\right)$$
$$+ 2K_w\alpha^2 \frac{dW}{d\xi}\left(\phi + 2\frac{dW}{d\xi}\right)$$
$$- 8K_g\alpha^2 \frac{d^3W}{d\xi^3}\left(\phi + 2\frac{dW}{d\xi}\right) + 4\hat{N}_\theta\left(\frac{dW}{d\xi}\right)^2$$
$$+ K_w W^2 - 4K_g W \frac{d^2W}{d\xi^2}. \tag{7.12}$$

Substituting Eqs. (3.14) and (3.15) in Eq. (7.12) and minimizing λ^2 with respect to the constant coefficients, the following eigenvalue value problem is obtained:

$$[K]\{Z\} = \lambda^2 [M_a]\{Z\}, \tag{7.13}$$

where Z is a column vector of constants.

Here stiffness matrix K and mass matrix M_a for TBT nanobeams are given as follows:

$$K = \begin{bmatrix} k_1 & k_2 \\ k_3 & k_4 \end{bmatrix},$$

where k_1, k_2, k_3, and k_4 are submatrices and are given as

$k_1(i, j) = \int_{-1}^{1} (8(\frac{1}{\Omega} + K_w\alpha^2 + \hat{N}_\theta)\varphi_i'\varphi_j' - 16(\hat{N}_\theta\alpha^2 + K_g\alpha^2)\varphi_i'''\varphi_j' - 16(\hat{N}_\theta\alpha^2 + K_g\alpha^2)\varphi_i'\varphi_j''' + 2K_w\varphi_i\varphi_j - 4K_g\varphi_i''\varphi_j - 4K_g\varphi_i\varphi_j'')d\xi,$

$k_2(i,j) = \int_{-1}^{1}(-8\hat{N}_\theta\alpha^2\varphi_i''\psi_j' + 2K_w\alpha^2\varphi_i\psi_j' - 8K_g\alpha^2\varphi_i''\psi_j' + 4\frac{1}{1/\Omega}$
$\varphi_i'\psi_j - (8\hat{N}_\theta\alpha^2 + 8K_g\alpha^2)\varphi_i'''\psi_j + 2K_w\alpha^2\varphi_i'\psi_j)d\xi,$
$k_3(i,j) = \int_{-1}^{1}(-8\hat{N}_\theta\alpha^2\psi_i'\varphi_j'' + 2K_w\alpha^2\psi_i'\varphi_j - 8K_g\alpha^2\psi_i'\varphi_j'' + 4\frac{1}{\Omega}\psi_i\varphi_j'$
$-8\hat{N}_\theta\alpha^2\psi_i\varphi_j''' + 2K_w\alpha^2\psi_i\varphi_j' - 8K_g\alpha^2\psi_i\varphi_j''')d\xi,$
$k_4(i,j) = \int_{-1}^{1}(8\psi_i'\psi_j' + 2\frac{1}{\Omega}\psi_i\psi_j)d\xi,$

$$M_a = \begin{bmatrix} m_1 & m_2 \\ m_3 & m_4 \end{bmatrix},$$

where m_1, m_2, m_3, and m_4 are submatrices and are defined as

$$m_1(i,j) = \int_{-1}^{1}(2\varphi_i\varphi_j + 8\alpha^2\varphi_i'\varphi_j')d\xi,$$

$$m_2(i,j) = \int_{-1}^{1}(2\alpha^2\varphi_i\psi_j' + 2\alpha^2\varphi_i'\psi_j)d\xi,$$

$$m_3(i,j) = \int_{-1}^{1}(2\alpha^2\psi_i\varphi_j' + 2\alpha^2\psi_i\varphi_j')d\xi,$$

$$m_4(i,j) = \int_{-1}^{1}(2\tau\psi_i\psi_j + 2\tau\alpha^2\psi_i'\psi_j')d\xi.$$

7.5 Reddy–Bickford Beam Theory (RBT)

Governing equations for vibration analysis of nanobeams embedded in elastic foundations are obtained as

$$-\rho A\omega^2 w_0 = G\tilde{A}\left(\frac{d\phi_0}{dx} + \frac{d^2w_0}{dx^2}\right) - \bar{N}\frac{d^2w_0}{dx^2} - k_w w_0 + k_g\frac{d^2w_0}{dx^2}$$
$$+ \mu\left[\bar{N}\frac{d^4w_0}{dx^4} + k_w\frac{d^2w_0}{dx^2} - k_g\frac{d^4w_0}{dx^4} - \rho A\omega^2\frac{d^2w_0}{dx^2}\right]$$
$$+ c_1EJ\frac{d^3\phi_0}{dx^3} - c_1^2EK\left(\frac{d^3\phi_0}{dx^3} + \frac{d^4w_0}{dx^4}\right), \quad (7.14)$$

$$E\hat{I}\frac{d^2\phi_0}{dx^2} - c_1E\hat{J}\left(\frac{d^2\phi_0}{dx^2} + \frac{d^3w_0}{dx^3}\right) - G\tilde{A}\left(\phi_0 + \frac{dw_0}{dx}\right) = 0. \quad (7.15)$$

Eliminating ϕ_0 from Eqs. (7.14) and (7.15), governing equations may be written as

$$-\left(\frac{68}{84}\bar{N}+\frac{105}{84EI}G\tilde{A}\mu\bar{N}\right)\frac{d^4w_0}{dx^4}+\frac{105}{84EI}G\tilde{A}\bar{N}\frac{d^2w_0}{dx^2}+\frac{68}{84}\mu\bar{N}\frac{d^6w_0}{dx^6}$$
$$=\rho A\omega^2\frac{105}{84EI}G\tilde{A}w_0-\frac{105}{84EI}G\tilde{A}k_w w_0$$
$$\left(\frac{68}{84}k_w+\frac{105}{84EI}G\tilde{A}k_g+\frac{105}{84EI}G\tilde{A}\mu k_w-\frac{68}{84}\rho A\omega^2\right.$$
$$\left.-\frac{105}{84EI}G\tilde{A}\mu\rho A\omega^2\right)\frac{d^2w_0}{dx^2}$$
$$-\left(\frac{68}{84}k_g+\frac{68}{84}\mu k_w+\frac{105}{84EI}G\tilde{A}\mu k_g+\frac{21}{84}G\tilde{A}-\frac{68}{84}\mu\rho A\omega^2\right)\frac{d^4w_0}{dx^4}$$
$$+\left(\frac{68}{84}\mu k_g+\frac{1}{105}EI\right)\frac{d^6w_0}{dx^6}. \tag{7.16}$$

Equation (7.16) may be written in non-dimensional form as

$$\lambda^2\left(\frac{105}{84}\bar{\Omega}W-\left(\frac{68}{84}+\frac{105}{84}\bar{\Omega}\alpha^2\right)\frac{d^2W}{dx^2}+\frac{68}{84}\alpha^2\frac{d^4W}{dx^4}\right)$$
$$=\left(-\frac{68}{84}K_w-\frac{105}{84}\bar{\Omega}K_g-\frac{105}{84}\bar{\Omega}\alpha^2K_w+\frac{105}{84}\bar{\Omega}\hat{N}_\theta\right)\frac{d^2W}{dX^2}$$
$$+\left(\frac{68}{84}K_g-\frac{68}{84}\hat{N}_\theta+\frac{68}{84}\alpha^2K_w+\frac{105}{84}\bar{\Omega}\alpha^2K_g+\frac{105}{84}\bar{\Omega}\right.$$
$$\left.-\frac{105}{84}\bar{\Omega}\alpha^2\hat{N}_\theta\right)\frac{d^4W}{dX^4}$$
$$+\left(-\frac{68}{84}\alpha^2K_g-\frac{1}{105}+\frac{68}{84}\alpha^2\hat{N}_\theta\right)\frac{d^6W}{dX^6}+\frac{105}{84}\bar{\Omega}K_w. \tag{7.17}$$

Application of DQM in Eq. (7.17), one may obtain generalized eigenvalue problem for RBT as

$$[K]\{W\}=\lambda^2[M_a]\{W\}, \tag{7.18}$$

where K is the stiffness matrix and M_a is the mass matrix.

7.6 Numerical Results and Discussions

Vibration of SWCNTs embedded in elastic medium under the influence of temperature has been investigated. The elastic medium is modeled as Winkler-type and Pasternak-type foundations. The effective properties of SWCNTs are taken as (Benzair *et al.* 2008; Murmu and Pradhan 2009b): $E = 1000$ GPa, $\nu = 0.19$, $\alpha_x = -1.6 \times 10^{-6}$ for room or low temperature, and $\alpha_x = 1.1 \times 10^{-6}$ for high temperature. A computer code is developed by the authors in MATLAB.

7.6.1 *Convergence*

First of all, convergence test has been performed to find minimum number of terms required for computing results. As such, Table 7.4 shows convergence of the first three frequency parameters ($\sqrt{\lambda}$) of EBT nanobeams with C–C support. In this table, we have taken $K_w = 50$, $K_g = 2$, $\theta = 20\ K$, $e_o a = 0.5$ nm, and $L/h = 10$. Similarly, Table 7.5 shows convergence test for TBT nanobeams. In this table, we have taken $K_w = 50$, $K_g = 2$, $\theta = 10\ K$, $e_0 a = 1$ nm, $L/h = 10$ with C–C edge condition. Figure 7.24 shows convergence of fundamental frequency parameter of nanobeams based on RBT where we have taken $K_w = 50$, $K_g = 4$, $\theta = 20\ K$, $e_0 a = 1$ nm, $L/h = 10$ with S–S edge condition. Above convergence patterns show that 10 grid points are sufficient to obtain results in the present analysis.

Table 7.4 Convergence of the first three frequency parameters of nanobeams (EBT)

n	First	Second	Third
3	4.8718	7.8096	10.7781
4	4.8718	7.7285	10.7781
5	4.8718	7.7285	10.4761
6	4.8718	7.7273	10.4761
7	4.8718	7.7273	10.4657
8	4.8718	7.7273	10.4657
9	4.8718	7.7273	10.4656
10	4.8718	7.7273	10.4656
11	4.8718	7.7273	10.4656

Table 7.5 Convergence of the first three frequency parameters (TBT)

n	First	Second	Third
3	3.6493	6.4704	13.9265
4	3.5402	6.4669	9.1435
5	3.5402	5.8616	9.1309
6	3.5399	5.8616	7.9217
7	3.5399	5.8511	7.9214
8	3.5399	5.8511	7.8697
9	3.5399	5.8511	7.8697

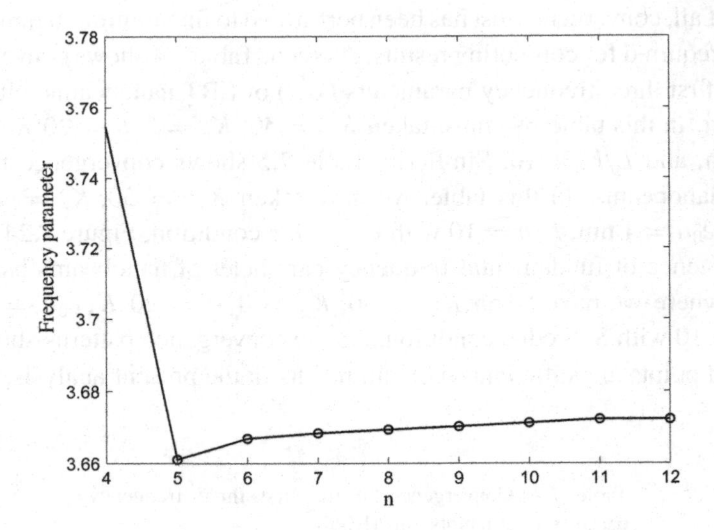

Fig. 7.24 Convergence of fundamental frequency parameter (RBT)

7.6.2 *Validation*

To validate the present results of Euler–Bernoulli, a comparison study has been carried out with the results of Wang *et al.* (2007). For this comparison, we have taken $K_w = 0$, $K_g = 0$, and $\theta = 0\ K$. As such, Figs. 7.25 and 7.26 show graphical comparisons of EBT and TBT nanobeams, respectively. In these figures, we have considered S–S support with $L/h = 10$. Similarly, tabular comparison study has been tabulated in Table 7.6 with that of Ansari and Sahmani (2011) for $L/h = 40$. Same parameters as that of

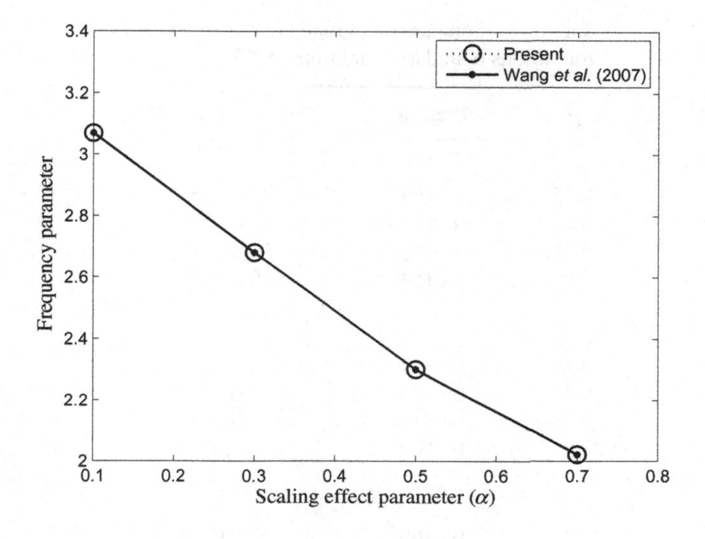

Fig. 7.25 Comparison of fundamental frequency parameter (EBT)

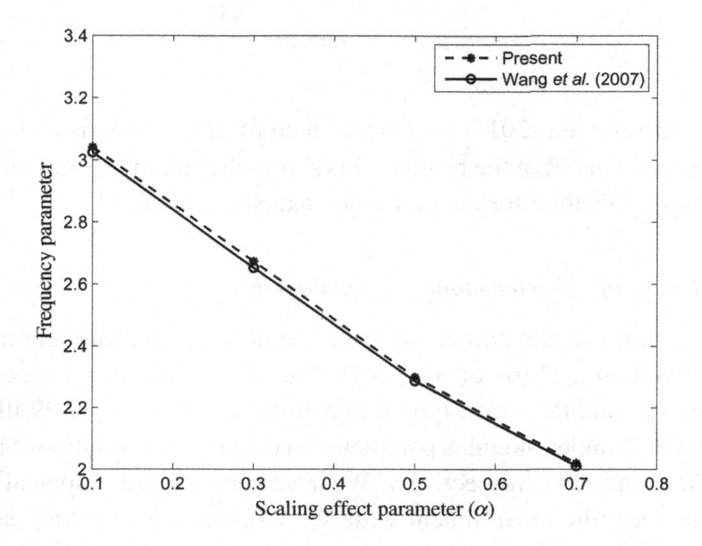

Fig. 7.26 Comparison of fundamental frequency parameter (TBT)

Table 7.6 Comparison of fundamental frequencies for various boundary conditions (RBT)

μ	Present	Amirian *et al.* (2013)
S–S		
0.25	0.0164	0.0231
0.5	0.0160	0.0231
0.75	0.0156	0.0231
1	0.0154	0.0.0231
C–S		
0.25	0.0321	0.0358
0.5	0.0320	0.0358
0.75	0.0310	0.0358
1	0.0309	0.0357
C–C		
0.25	0.0468	0.0524
0.5	0.0464	0.0523
0.75	0.0460	0.0522
1	0.0459	0.0521

Ansari and Sahmani (2011) have been taken for this comparison. One may find close agreement of the results. This shows the suitability and reliability of the present method for the vibration analysis of SWCNTs.

7.6.3 *Effect of Winkler modulus parameter*

In this section, we have investigated the influence of surrounding medium on the vibration analysis of SWCNTs. The elastic medium is modeled as Winkler-type and Pasternak-type foundations. Figures 7.27–7.29 illustrate the effect of Winkler modulus parameter on the vibration solutions based on EBT, TBT, and RBT, respectively. We have shown these graphical results in low-temperature environment with $K_g = 0$. Numerical values taken for this computation are $\theta = 20\ K$, $L/h = 10$ in Fig. 7.27 with C–S support, whereas in Fig. 7.28, we have taken $\theta = 20\ K$, $L/h = 10$ with C–S support and in Fig. 7.29, we have taken $\theta = 20\ K$, $L/h = 20$ with C–S support. The Winkler modulus parameter is taken in the range of 0–400. It is observed from these figures that the frequency parameter associated with the first

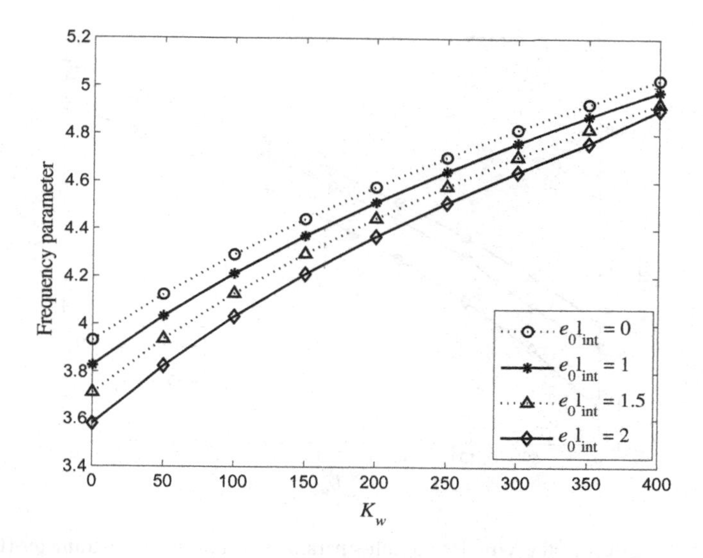

Fig. 7.27 Effect of the Winkler modulus parameter on frequency parameter (EBT)

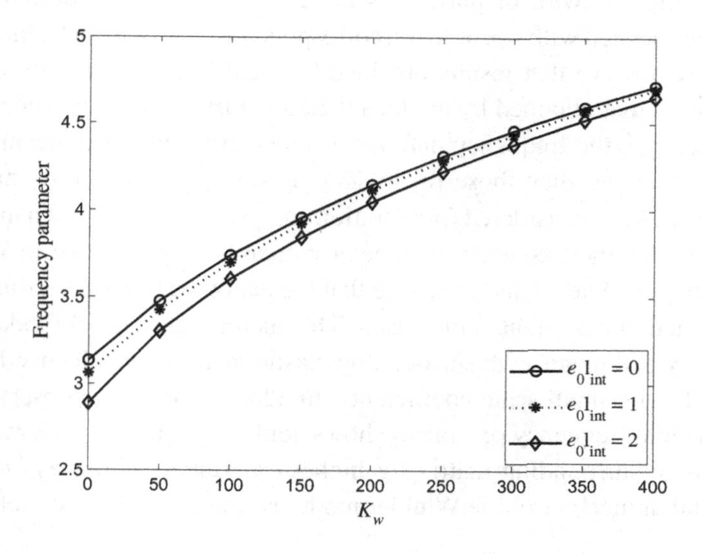

Fig. 7.28 Effect of the Winkler modulus parameter on frequency parameter (TBT)

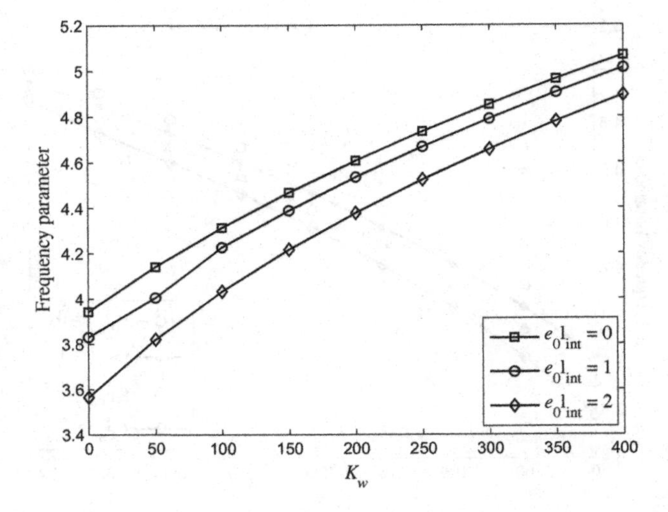

Fig. 7.29 Effect of the Winkler modulus parameter on frequency parameter (RBT)

mode decreases with increase in scale coefficient. It may be noted that results associated with $e_0a = 0$ nm correspond to those of local beam theory. One may observe that results obtained by local beam theory are overpredicted than that obtained by nonlocal beam theory. As the scale coefficient e_0a increases, the frequency parameter obtained for nonlocal beam theory becomes smaller than those for its local counterpart. Therefore, nonlocal theory should be considered for vibration analysis of structures at nanoscale. It is seen that the frequency parameter increases with increase in Winkler modulus parameter. This is because that the nanotube becomes stiffer when elastic medium constant is increased. This increasing trend of fundamental frequency parameter with surrounding elastic medium is influenced significantly by the small scale coefficients. In addition, it is also observed that fundamental frequency parameter shows nonlinear behavior with respect to stiffness of surrounding matrix for higher e_0a values. This may be due to the fact that increase of the Winkler modulus causes CNT to be more rigid.

7.6.4 *Effect of Pasternak shear modulus parameter*

In this section, effect of Pasternak shear modulus parameter on the vibration analysis has been examined. Figures 7.30–7.32 show the distribution of

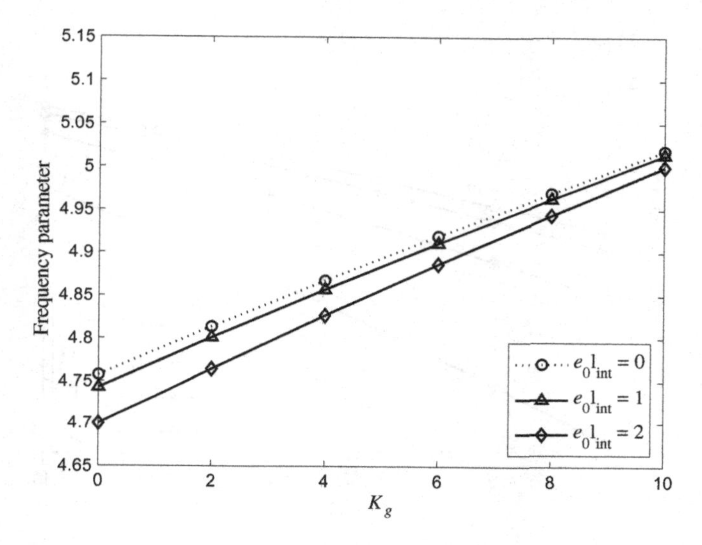

Fig. 7.30 Effect of the shear modulus parameter on frequency parameter (EBT)

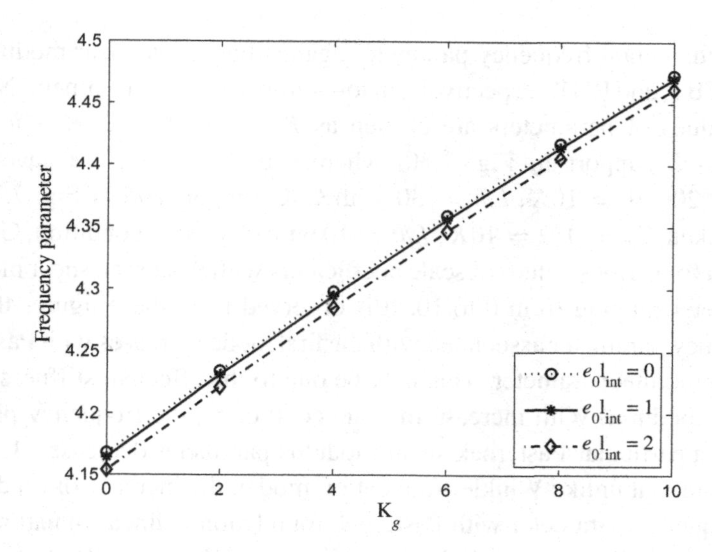

Fig. 7.31 Effect of the shear modulus parameter on frequency parameter (TBT)

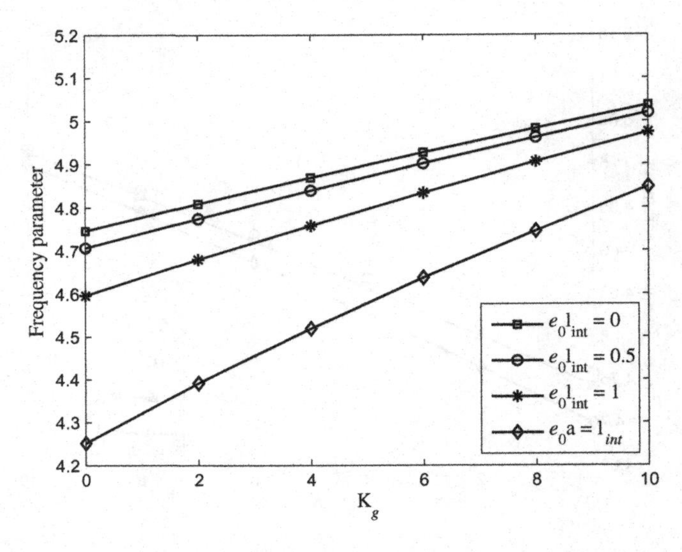

Fig. 7.32 Effect of the shear modulus parameter on frequency parameter (RBT)

the fundamental frequency parameter against Pasternak shear modulus for EBT, TBT, and RBT, respectively, in low-temperature environment. Numerical values of parameters are chosen as $K_w = 0, \theta = 20K, L/h = 30$ with C–C support in Fig. 7.30, whereas in Fig. 7.31, we have taken $K_w = 200, \theta = 10K, L/h = 30$ with C–C support and in Fig. 7.32, we have taken $K_w = 0, \theta = 10K, L/h = 10$ with C–C edge condition. Graph is plotted for various values of scale coefficients with Pasternak shear modulus parameter ranging from 0 to 10. It is observed from these figures that the frequency parameter associated with the first mode increases with Pasternak shear modulus parameter. This may be due to the effective stiffness of the elastic medium. With increase in scale coefficient, the frequency parameter for a particular Pasternak shear modulus parameter decreases. It is also observed that unlike Winkler foundation model, the increase of fundamental frequency parameter with Pasternak foundation is linear in nature. This is due to the domination of elastic medium modeled as the Pasternak-type foundation model. Same observation has also been reported in Murmu and Pradhan (2009b). Next, we have analyzed the effect of Pasternak foundation model over Winkler foundation model. As such, Fig. 7.33 illustrates the fundamental frequency parameter of RBT nanobeams as a function of small

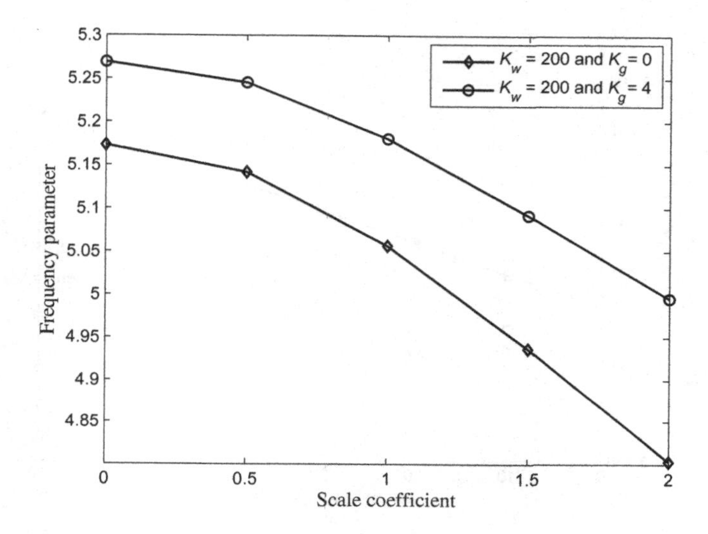

Fig. 7.33 Effect of the frequency parameter with small scale coefficients

scale coefficient in low-temperature environment with $L/h = 10$ and C–C edge condition. It may be observed that the frequency parameter obtained from Pasternak foundation model is relatively larger than those obtained from the Winkler foundation model.

7.6.5 *Effect of temperature*

Effect of temperature on the vibration of nanobeams embedded in an elastic medium has been investigated. Effect of temperature on the vibration solutions has been illustrated in Figs. 7.34–7.36, respectively for EBT, TBT, and RBT. In Fig. 7.34, we have taken S–S nanobeams with $L/h = 10$, $e_0a = 1.5$ nm, $K_w = 10$, $K_g = 4$. Similarly, we have taken $L/h = 50$, $e_0a = 0.5$ nm, $K_w = 50$, $K_g = 2$ with C–C support in Fig. 7.35 and $L/h = 30$, $e_0a = 1$ nm, $K_w = 50$, $K_g = 2$ with C–C support in Fig. 7.36. It is noticed that the fundamental frequency parameter increases with increase in temperature in low-temperature environment, while they decrease with increase in temperature in high-temperature environment. Thus, one may say that the fundamental frequency parameter considering thermal effect are larger than those ignoring the influence of temperature change in low-temperature environment. While in high-temperature environment, the fundamental frequency

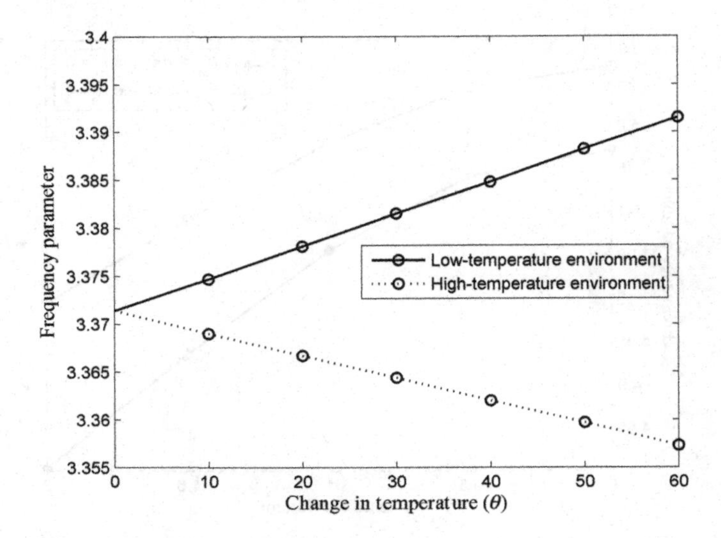

Fig. 7.34 Change of frequency parameter with temperature change (EBT)

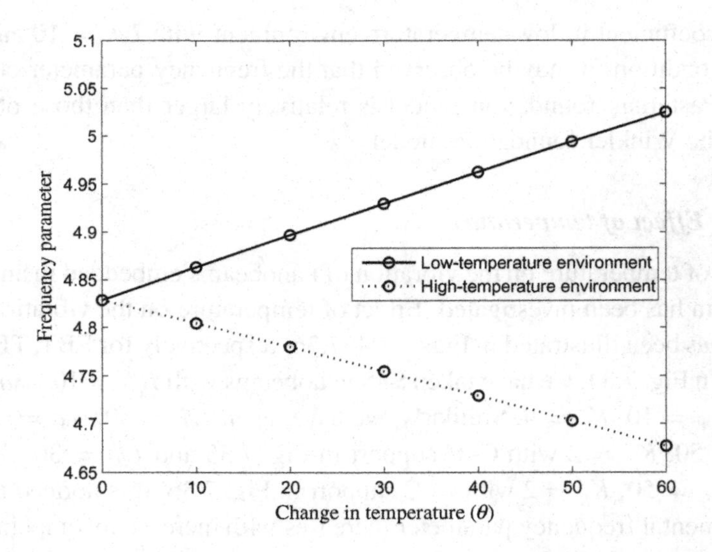

Fig. 7.35 Change of frequency parameter with temperature change (TBT)

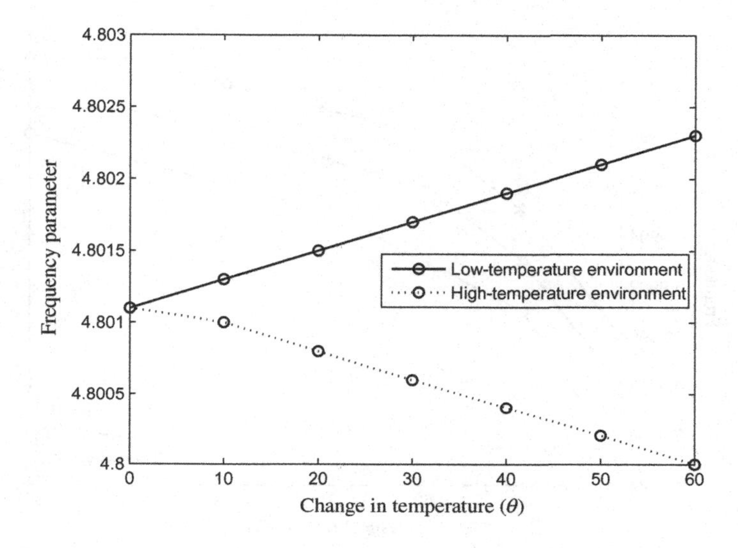

Fig. 7.36 Change of frequency parameter with temperature change (RBT)

parameter considering thermal effect are smaller than those excluding the influence of temperature change. Same observation have also been noted in Murmu and Pradhan (2009a, 2010), Zidour *et al.* (2012), Maachou *et al.* (2011).

7.6.6 *Effect of aspect ratio*

To illustrate the effect of aspect ratio on the fundamental frequency parameter, variation of frequency ratio with the aspect ratio (L/h) has been shown in Fig. 7.37 for different magnitudes of temperature change. Results have been shown for EBT nanobeams with $K_w = 50$, $K_g = 2$, $e_0a = 1$ nm^2, and C–C support in low-temperature environment. It is observed that the frequency ratio increases with increase in aspect ratio. In addition, it is also seen that the fundamental frequency parameter is also dependent on temperature change (θ). The differences in magnitudes of frequency parameter for different temperature changes are larger in low aspect ratios, while the differences in magnitudes of frequency parameter for different temperature changes are smaller for large aspect ratios.

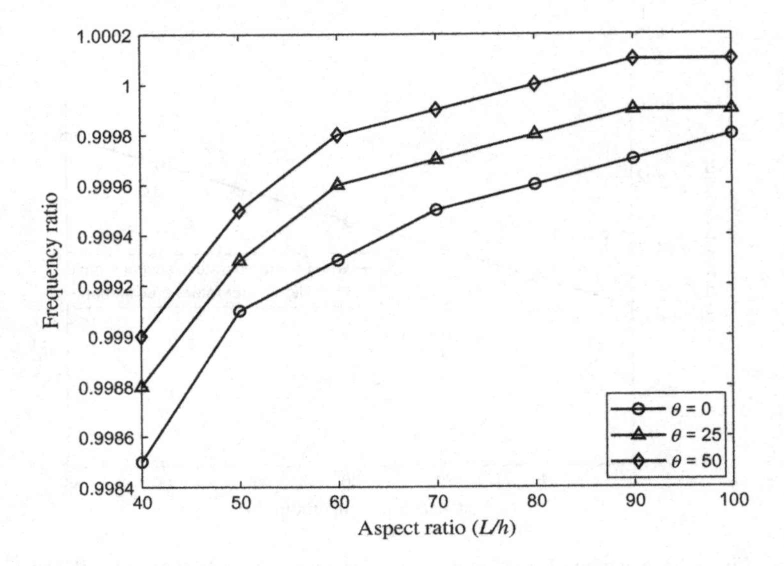

Fig. 7.37 Change of frequency ratio with aspect ratio

7.7 Conclusions

This study includes the vibration characteristics of free vibration of non-uniform nanobeams based on nonlocal elasticity theory. Boundary characteristic orthogonal polynomials have been generated and are applied in the Rayleigh–Ritz method to study the effect of non-uniform parameters on the frequency parameters of Euler–Bernoulli and Timoshenko nanobeams. Non-uniform material properties of nanobeams are assumed taking linear and quadratic variations of Young's modulus and density. Effects of non-uniform parameter, boundary condition, aspect ratio, and scaling effect parameter on the frequency parameters have been investigated. The first four deflection shapes are given for S–S and C–S boundary conditions with scaling effect parameters as 0 and 0.3.

Again, Rayleigh–Ritz method has been applied to investigate the thermal effect on the vibration of nanobeams embedded in elastic medium based on EBT and TBT, whereas DQM has been applied to investigate the thermal effect on the vibration of nanobeams embedded in an elastic medium based on nonlocal RBT. Theoretical formulations include effects of small scale, elastic medium, and temperature change. It is seen that results obtained based on local beam theory are overestimated. The frequency parameter increases with increase in temperature and Winkler and Pasternak coefficients of elastic foundation.

Chapter 8

Bending and Buckling of Nanoplates

In this chapter, we have discussed bending and buckling of nanoplates based on classical plate theory (CPT) in conjunction with nonlocal elasticity theory of Eringen. Two-dimensional simple polynomials have been used as shape functions in the Rayleigh–Ritz method. Complicating effects such as Winkler and Pasternak foundation models have been considered.

8.1 Bending of Nanoplates

We have investigated bending of nanoplates in the absence of elastic foundation. As such, the strain energy may be obtained from Eq. (2.28) by setting $k_w = k_p = 0$.

$$U = \frac{1}{2}D\int_0^a\int_0^b\left\{\left(\frac{\partial^2 w}{\partial x^2}\right)^2 + 2\nu\left(\frac{\partial^2 w}{\partial x^2}\frac{\partial^2 w}{\partial y^2}\right) + \left(\frac{\partial^2 w}{\partial y^2}\right)^2 + 2(1-\nu)\left(\frac{\partial^2 w}{\partial x \partial y}\right)^2\right\}dx\,dy, \tag{8.1}$$

where $R = a/b$ is the aspect ratio.

Similarly, the potential energy of the transverse force q may be given by Eq. (3.29).

We have introduced the non-dimensional variables $X = x/a$ and $Y = y/b$.

The total potential energy U_T of the system may be written in non-dimensional form as

$$U_T = \frac{1}{2}\int_0^1\int_0^1 \left\{ \frac{D}{a^4}\left[\left(\frac{\partial^2 w}{\partial X^2}\right)^2 + 2\nu R^2 \left(\frac{\partial^2 w}{\partial X^2}\frac{\partial^2 w}{\partial Y^2}\right) + R^4\left(\frac{\partial^2 w}{\partial Y^2}\right)^2 + 2(1-\nu)R^2\left(\frac{\partial^2 w}{\partial X \partial Y}\right)^2 \right] - q\left[w - \frac{\mu}{a^2}\left(\frac{\partial^2 w}{\partial X^2} + R^2\frac{\partial^2 w}{\partial Y^2}\right)\right] \right\} dX\, dY. \tag{8.2}$$

Substituting Eq. (3.31) in Eq. (8.2) and then minimizing U_T as a function of constants, one may obtain the following system of linear equation:

$$\sum_{j=1}^{n} a_{ij} c_j = P_c b_i, \tag{8.3}$$

where

$$a_{ij} = \int_0^1\int_0^1 \left[\phi_i^{XX}\phi_j^{XX} + R^4\phi_i^{YY}\phi_j^{YY} + \nu R^2\left(\phi_i^{XX}\phi_j^{YY} + \phi_i^{YY}\phi_j^{XX}\right) + 2(1-\nu)R^2\phi_i^{XY}\phi_j^{XY} \right] dX\, dY,$$

$$b_i = \int_0^1\int_0^1 \left(\phi_i - \frac{\mu}{a^2}\left(\phi_i^{XX} + R^2\phi_i^{YY}\right)\right) dX\, dY,$$

and

$$P_c = \frac{qa^4}{D}.$$

8.2 Numerical Results and Discussions

Above system of linear equation has been solved by using MATLAB. It should be noted that the present study is subjected to uniformly distributed loading ($q(X) = q_0$). Non-dimensional maximum center deflection ($W_{\max}$) is given by $-w \times (Eh^2/q_0 a^4) \times 100$ Aghababaei and Reddy (2009).

8.2.1 *Convergence*

First of all, convergency has been carried out to find the minimum number of terms required for computation. As such, Fig. 8.1 shows convergency of

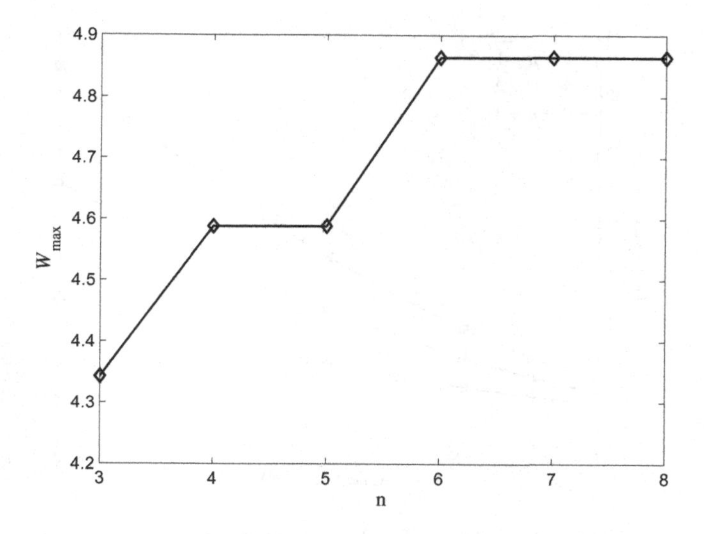

Fig. 8.1 Variation of non-dimensional maximum center deflection ($W_{\max}$) with n

Table 8.1 Comparison of non-dimensional maximum center deflection of S–S–S–S nanoplate

	$\mu = 0$	$\mu = 0.5$	$\mu = 1$	$\mu = 1.5$	$\mu = 2$	$\mu = 2.5$	$\mu = 3$
Present	4.0673	4.3637	4.7601	5.0565	5.4530	5.8094	6.1058
Ref.*	4.0083	4.3702	4.7322	5.0942	5.4561	5.8181	6.1800

*Aghababaei and Reddy (2009).

simply supported nanoplates with $\mu = 0.5$ nm^2, $R = 1$, and $a = 10$ nm. One may find from this figure that convergency is attained at $n = 7$.

8.2.2 *Validation*

Next, to show the reliability and accuracy of the present method, our numerical results are compared with those available in literature for simply supported nanoplates. As such, tabular comparison has been given in Table 8.1 with Aghababaei and Reddy (2009). For this comparison, same parameters as that of Aghababaei and Reddy (2009) have been used. One can see that our results are in a good agreement with analytical solutions. It is noted that S–S–S–S would denote simply supported–simply supported–simply supported–simply supported boundary condition.

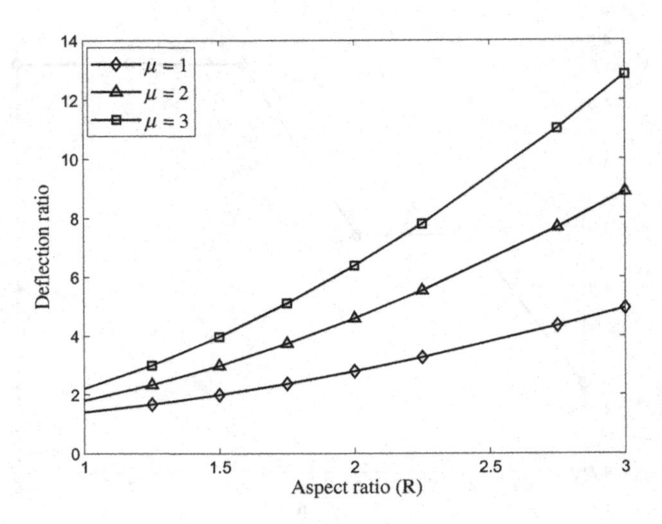

Fig. 8.2 Variation of deflection ratio with aspect ratio

In the following paragraphs, we have investigated some of the parametric studies.

8.2.3 *Effect of aspect ratio*

To investigate the effect of aspect ratio on the deflection, variation of deflection ratio with aspect ratio has been illustrated in Fig. 8.2 for F–S–F–S nanoplate with $a = 5$ nm. Here, we define deflection ratio as

$$\text{deflection ratio} = \frac{\text{deflection calculated by nonlocal theory}}{\text{deflection calculated using local theory}}.$$

One may notice that the deflection ratio increases with increase in aspect ratio. In other words, increasing aspect ratio will decrease non-dimensional maximum center deflection. Same observation has also been reported in Aghababaei and Reddy (2009).

8.2.4 *Effect of length*

Figure 8.3 depicts the effect of length on the non-dimensional maximum center deflection. Results have been shown for different nonlocal parameters with $R = 1$ and S–C–S–C edge condition. It is figured out that non-dimensional maximum center deflection decreases with increase in length.

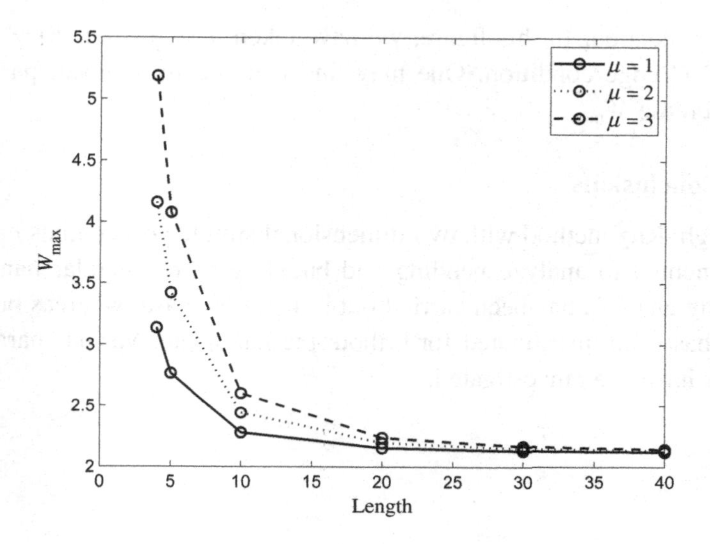

Fig. 8.3 Variation of W_{max} with length

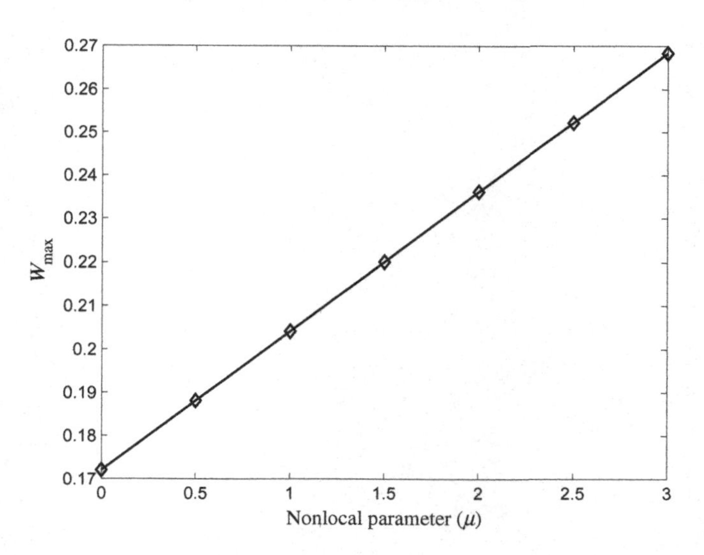

Fig. 8.4 Variation of W_{max} with nonlocal parameter

8.2.5 *Effect of nonlocal parameter*

To illustrate the effect of nonlocal parameter on the deflection, Fig. 8.4 illustrates the variation of nonlocal parameter with non-dimensional maximum center deflection. In this figure, we have taken $a = 10$ nm, $R = 2$ with F–C–F–C edge condition. One may find increase in nonlocal parameter will increase $W_{\max}$.

8.3 Conclusions

Rayleigh–Ritz method with two-dimensional simple polynomials has been implemented to analyze bending and buckling of rectangular nanoplate. Bending analysis has been carried out in isotropic case, whereas buckling study has been investigated for orthotropic nanoplate. Various parametric studies have been investigated.

Chapter 9

Vibration of Nanoplates

Here, we have considered vibration of isotropic nanoplate. The Rayleigh–Ritz method has been employed with two-dimensional simple polynomials as shape functions. Parametric studies such as effect of length, aspect ratio, and nonlocal parameter have been analyzed. Three-dimensional mode shapes for some specified boundary condition have been presented. Some of the important aspects have been highlighted.

Maximum strain energy may be obtained by setting k_w and k_p as zero in Eq. (3.35):

$$U = \frac{1}{2}D \int_0^a \int_0^b \left\{ \left(\frac{\partial^2 w_0}{\partial x^2} \right)^2 + 2\nu \left(\frac{\partial^2 w_0}{\partial x^2} \frac{\partial^2 w_0}{\partial y^2} \right) + \left(\frac{\partial^2 w_0}{\partial y^2} \right)^2 + 2(1-\nu) \left(\frac{\partial^2 w_0}{\partial x \partial y} \right)^2 \right\} dx\, dy. \tag{9.1}$$

Maximum kinetic energy may be written as Eq. (3.36).

Equating maximum kinetic and strain energies, one may obtain the Rayleigh quotient as

$$\frac{\rho h \omega^2}{D} = \frac{\int_0^a \int_0^b \left[\left(\frac{\partial^2 w_0}{\partial x^2} \right)^2 + 2\nu \left(\frac{\partial^2 w_0}{\partial x^2} \frac{\partial^2 w_0}{\partial y^2} \right) + \left(\frac{\partial^2 w_0}{\partial y^2} \right)^2 + 2(1-\nu) \left(\frac{\partial^2 w_0}{\partial x \partial y} \right)^2 \right] dx\, dy}{\int_0^a \int_0^b \left[w_0^2 + \mu \left(\left(\frac{\partial w_0}{\partial x} \right)^2 + \left(\frac{\partial w_0}{\partial y} \right)^2 \right) \right] dx\, dy}, \tag{9.2}$$

where m_0 is taken as ρh.

We have introduced the non-dimensional variables $X = x/a$ and $Y = y/b$.

As such, non-dimensional Rayleigh quotient is obtained as

$$\frac{\rho h\omega^2 a^4}{D}=\frac{\int_0^1\int_0^1\left[\left(\frac{\partial^2 W}{\partial X^2}\right)^2+2\nu R^2\left(\frac{\partial^2 W}{\partial X^2}\frac{\partial^2 W}{\partial Y^2}\right)+R^4\left(\frac{\partial^2 W}{\partial Y^2}\right)^2+2(1-\nu)R^2\left(\frac{\partial^2 W}{\partial X\partial Y}\right)^2\right]dX\,dY}{\int_0^1\int_0^1\left[W^2+\frac{\mu}{a^2}\left(\left(\frac{\partial W}{\partial X}\right)^2+R^2\left(\frac{\partial W}{\partial Y}\right)^2\right)\right]dX\,dY}. \tag{9.3}$$

Substituting Eq. (3.31) into Eq. (9.3), we get a generalized eigenvalue problem as

$$[K]\{Z\}=\lambda^2[M]\{Z\}, \tag{9.4}$$

where $\lambda^2=\rho h a^4\omega^2/D$ is the non-dimensional frequency parameter, Z is a column vector of constants, and K and M are the stiffness and mass matrices, respectively, which are given as follows:

$$K_{ij}=\beta_{ij}^{2020}+\nu R^2\left(\beta_{ij}^{2002}+\beta_{ij}^{0220}\right)+R^4\beta_{ij}^{0202}+2(1-\nu)R^2\beta_{ij}^{1111},$$
$$M_{ij}=\beta_{ij}^{0000}+\mu\left(\frac{1}{a}\right)^2\left(\beta_{ij}^{1010}+R^2\beta_{ij}^{0101}\right),$$

where

$$\beta_{ij}^{klmn_a}=\int_0^1\int_0^1\left[\frac{\partial^{k+l}\phi_i}{\partial X^k\partial Y^l}\right]\left[\frac{\partial^{m+n_a}\phi_j}{\partial X^m\partial Y^{n_a}}\right]dX\,dY.$$

9.1 Numerical Results and Discussions

Numerical values of the frequency parameter λ have been obtained by solving Eq. (9.4) using a computer program developed by the authors in MATLAB which is run for different values of n to get appropriate value of the order of approximation n.

9.1.1 *Convergence*

Table 9.1 shows convergence of first three frequency parameters for S–S–S–S, S–C–S–C, F–C–F–C, and F–S–F–S nanoplates with $\nu=0.3$, $R=2$, $\mu=2\ \text{nm}^2$, and $a=10$ nm. From this table, it can be clearly seen that the frequency parameters (λ) approach to the results with the increasing value of n and further increase of n does not have any effect on the results.

Table 9.1 Convergence of the first three frequency parameters of nanoplates

	S–S–S–S			S–C–S–C		
n	First	Second	Third	First	Second	Third
10	35.0202	49.2575	80.6866	64.3936	69.6776	93.8351
15	35.0086	49.2575	68.5520	64.3898	69.6776	81.3358
20	35.0086	49.1652	68.5520	64.3898	69.6363	81.3358
25	35.0086	49.1652	67.9560	64.3898	69.6363	80.9277
30	35.0086	49.1650	67.9538	64.3898	69.6362	80.9271
35	35.0086	49.1646	67.9538	64.3898	69.6361	80.9271
40	35.0086	49.1646	67.9435	64.3898	69.6361	80.9210
45	35.0086	49.1646	67.9433	64.3898	69.6361	80.9210
47	35.0086	49.1646	67.9433	64.3898	69.6361	80.9210
	F–C–F–C			F–S–F–S		
	First	Second	Third	First	Second	Third
10	61.8442	62.2421	64.7394	29.2401	32.9761	44.9044
15	61.8442	61.9872	64.6373	29.0358	32.9761	42.9748
20	61.8388	61.9872	64.6373	29.0358	32.7816	42.9748
25	61.8388	61.9860	64.5829	29.0261	32.7816	42.7648
30	61.8385	61.9855	64.5823	29.0260	32.7780	42.7610
35	61.8255	61.9855	64.5823	29.0260	32.7769	42.7610
40	61.8255	61.9324	64.5823	29.0259	32.7769	42.7595
45	61.7567	61.9324	64.5815	29.0259	32.7769	42.7595
47	61.7567	61.9324	64.5815	29.0259	32.7769	42.7595

9.1.2 *Validation*

For comparison of the results with analytical solutions of Aksencer and Aydogdu (2011), numerical results have been given in graphical form in Fig. 9.1 where variation of the first two frequency ratios with length a is given for nanoplates with S–S–S–S boundary condition. Here the aspect ratio is taken as 1 and nonlocal parameters are taken as 0, 1, 2, and 4 nm^2. From this figure, we may say that increase in nonlocal parameter decreases frequency ratios. This decreasing behavior is clearly noticed for higher modes. This is because of the fact that increase in mode number decreases the wavelength. Since nonlocal effects are more pronounced for smaller wavelengths, there will be no nonlocal effects after a certain length of nanoplates. All these observations are expected and a close agreement of the results with Aksencer and Aydogdu (2011) is seen.

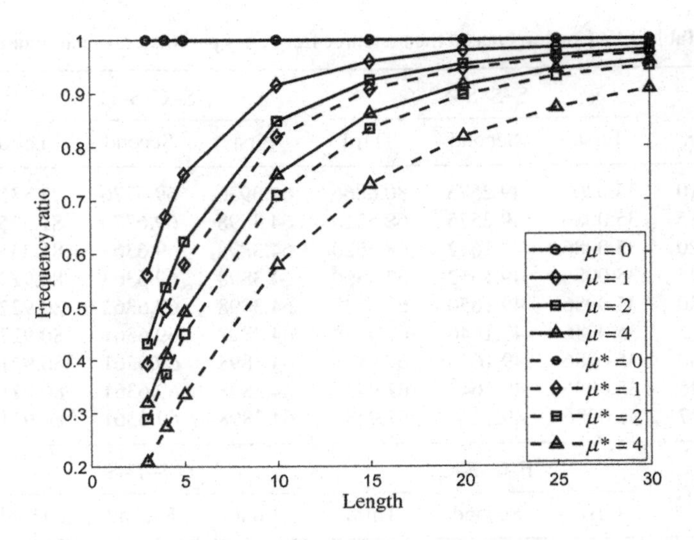

Fig. 9.1 Variation of frequency ratios with length of nanoplates

9.1.3 *Frequency parameters for different boundary conditions*

The first four frequency parameters are tabulated in Table 9.2 for nanoplates subjected to boundary conditions (S–S–S–S, C–C–C–C, S–C–S–C, F–C–F–C, and F–S–F–S) with nonlocal parameters (0, 1, 2, and 4 nm^2), length (a = 5 nm), and aspect ratio as 1. From this table, it may be seen that frequency parameters are highest in C–C–C–C and lowest in F–S–F–S for a particular value of nonlocal parameter. Frequency parameters decrease with increase in nonlocal parameter in all the boundary conditions. Table 9.3 gives the first four frequency parameters for nanoplates subjected to various boundary conditions having $\mu = 2$ nm^2, aspect ratio = 2, and length = 5 nm. From Tables 9.2 ($\mu = 2$ nm^2) and 9.3, it is observed that the frequency parameters increase with increase in aspect ratio for a particular boundary condition and nonlocal parameter. It is also seen that frequency parameters are highest in C–C–C–C than other boundary conditions.

9.1.4 *Effect of aspect ratio*

To see the influence of aspect ratio on the frequency parameters of nanoplates, the behavior of fundamental frequency parameter of C–C–C–C nanoplates with aspect ratio is plotted in Fig. 9.2 for length = 10 nm.

Table 9.2 First four frequency parameters of nanoplates for $a = 5$ nm and $a/b = 1$

Mode No.	$\mu = 0$	$\mu = 1$	$\mu = 2$	$\mu = 4$
S–S–S–S				
1	19.7000	14.7556	12.2912	9.6800
2	49.3000	28.6157	22.1851	16.5455
3	49.3000	28.6157	22.1851	16.5455
4	79	38.7198	29.1902	21.3842
C–C–C–C				
1	36	25.6182	20.9293	16.2072
2	73.4000	40.2819	30.8650	22.8363
3	73.4000	40.2819	30.8650	22.8363
4	108.2000	50.2722	37.5982	27.4081
S–C–S–C				
1	29	21.1091	17.4090	13.5914
2	54.7000	31.3612	24.2491	18.0520
3	69.3000	38.4787	29.5537	21.9013
4	94.6000	45.1134	33.8697	24.7486
F–C–F–C				
1	22.2000	18.0585	15.2936	11.7908
2	26.5000	18.7871	15.5360	12.4295
3	43.6000	23.5071	18.0136	13.5242
4	61.2000	33.3133	24.8433	18.1651
F–S–F–S				
1	9.6000	8.1375	7.1737	5.9649
2	16.1000	11.7857	9.7195	7.5853
3	36.7000	20.0205	15.3247	11.3391
4	38.9000	24.1377	18.9747	14.2818

The graph is shown for different nonlocal parameters. From the above graph, it is seen that nonlocal effect on frequency parameters is more prominent in greater values of aspect ratio. This may be due to the fact that when the aspect ratio increases, nanoplates become smaller for a specified length of

Table 9.3 First four frequency parameters for $a = 5$ nm, $\mu = 2$ nm^2, and $a/b = 2$

B.Cs	First	Second	Third	Fourth
S–C–S–C	39.6000	40.7306	45.2329	52.5002
S–C–S–S	29.5537	33.8697	41.0402	49.9379
S–C–S–F	12.6602	21.6101	29.8275	32.1331
S–S–S–F	9.7195	20.5617	23.4617	31.2146
S–F–S–F	7.0937	11.5810	18.7576	22.7213
C–C–C–C	40.2856	43.4408	50.1693	58.9610
C–C–C–S	30.8647	37.5978	46.7047	53.5259
C–C–S–S	30.0688	35.5235	43.6951	53.1091
C–C–F–F	10.3208	16.5303	24.7089	28.6041
C–F–C–F	14.9166	15.5252	25.8441	27.5855
C–F–S–F	10.8033	13.0646	23.1535	25.2039
C–F–F–F	2.9265	6.5858	11.3837	18.3020
S–S–F–F	4.5285	12.6233	19.9921	22.1292
S–F–F–F	5.8101	8.2758	16.5722	18.8076
F–F–F–F	9.4809	11.2065	18.9165	20.3403
S–F–S–C	12.6602	21.6101	29.8275	32.1331
S–C–C–C	39.7909	41.9053	47.5163	55.5969
S–C–C–S	30.0688	35.5235	43.6951	53.1091
C–F–F–S	5.7331	14.4925	20.8909	24.3087
C–F–S–C	14.2968	24.6622	30.2733	35.5735
C–F–S–S	12.1174	23.6980	24.4128	32.6778
S–S–S–S	22.1851	29.1902	38.2287	44.1800
C–C–C–F	16.6358	28.1756	30.7702	38.0487
C–S–C–F	15.2788	24.8379	27.8202	34.5728

the nanoplates. This leads to an increase in the small scale effect because size dependency plays a vital role in the nonlocal elasticity theory. Let us define the relative error percent as

$$\text{Relative error percent} = \frac{|\text{ Local result} - \text{nonlocal result }|}{|\text{ Local result}|} \times 100.$$

Neglecting nonlocal effect, the relative error percents for aspect ratios 1 and 3 with $\mu = 3$ nm^2 are 23.97% and 54.09%, respectively. Therefore, nonlocal theory should be considered for free vibration of small scaled rectangular nanoplates with high aspect ratios. It is also observed that for a particular nonlocal parameter, frequency parameters increase with increase in aspect ratio. When we compare all nonlocal parameters, frequency parameters are highest in the case of $\mu = 0$, i.e. local frequency parameter.

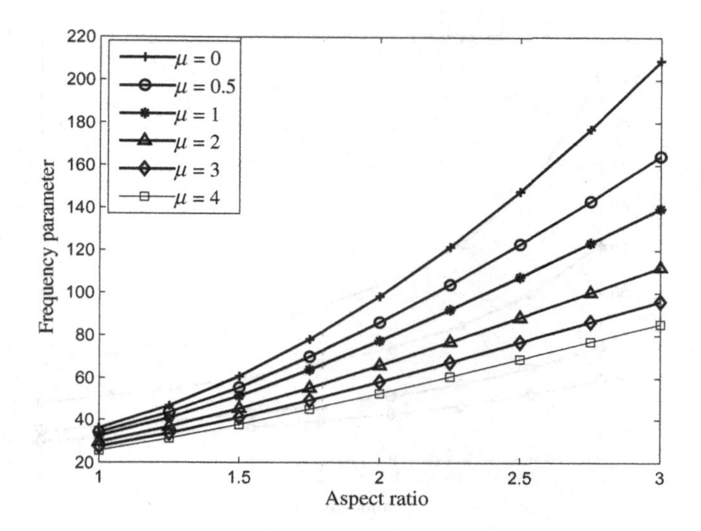

Fig. 9.2 Variation of fundamental frequency parameter with aspect ratio

This shows that frequency parameters are overpredicted when local beam model is considered for vibration analysis of nanoplates. As the scaling effect parameter increases, frequency parameters for nonlocal nanoplates become smaller than those of its local counterpart. This reduction can be clearly seen when we consider higher vibration modes. The reduction is due to the fact that the nonlocal model may be viewed as atoms linked by elastic springs, whereas in the case of local continuum model, the spring constant is assumed to take an infinite value. So small scale effect makes the nanoplates more flexible and nonlocal impact cannot be neglected. As such, nonlocal theory should be used for better predictions of high natural frequency of nanoplates.

9.1.5 *Effect of nonlocal parameter*

To investigate the effect of small scale in different vibration modes, non-dimensional frequency parameters associated with the first four modes are illustrated in Fig. 9.3 for different nonlocal parameters. Here we have considered F–S–F–S nanoplates with aspect ratio as 2 and length as 10 nm. It is noticed that frequency parameters decrease with nonlocal parameter in

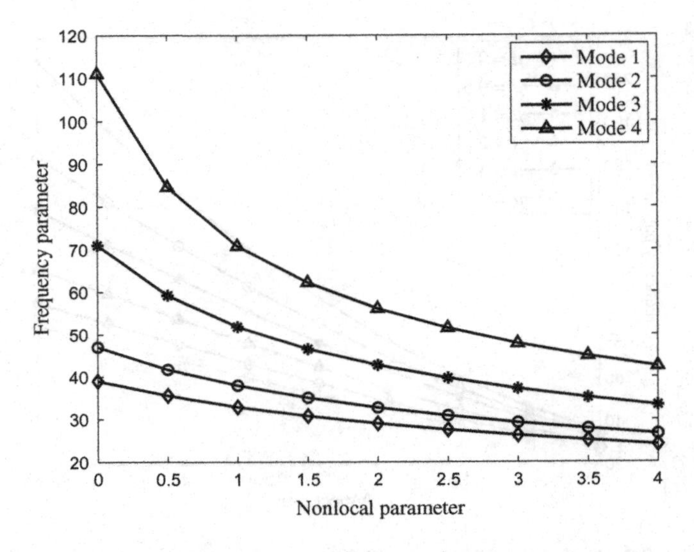

Fig. 9.3 Variation of the first four frequency parameters with nonlocal parameter

all the modes. It can be noted that this behavior is seen in all the boundary conditions. The nonlocal effect is more pronounced in higher vibration modes. Neglecting nonlocal effect, the relative error percents for the first and fourth mode number with $\mu = 2$ nm^2 are found to be 25.57% and 49.48%, respectively. This shows noticeable influence of the nonlocal parameter on the frequency parameters in higher modes.

9.1.6 *Effect of length*

Variation of the length of nanoplates on the fundamental frequency parameter of F–C–F–C nanoplates with $R = 2$ is shown in Fig. 9.4 for different nonlocal parameters. From this figure, it is observed that frequency parameters obtained by local elasticity theory (with non-zero nonlocal parameter) are always larger than those obtained by nonlocal theory of elasticity. It is also noticed that for each length of the nanoplates, frequency parameters decrease with increase in the nonlocal parameter. Again, one may see that as the length of nanoplates increases, frequency parameters increase for each value of the nonlocal parameter. This is due to the fact that size dependency plays a vital role in the nonlocal elasticity theory. In other words, by increasing the length of the nanoplates (a) and assuming l_{int} to be unchanged,

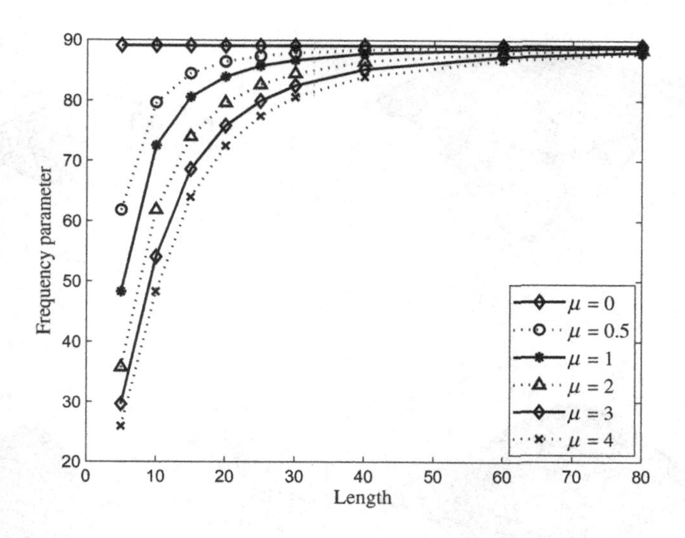

Fig. 9.4 Variation of fundamental frequency parameter with length

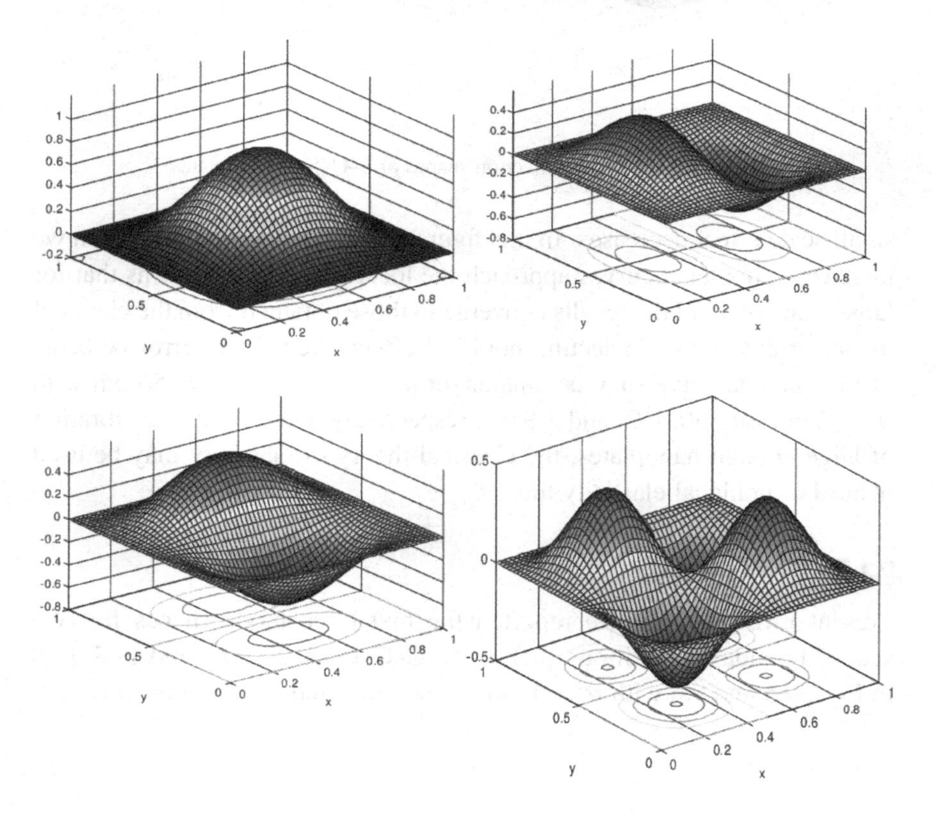

Fig. 9.5 First four deflection shapes of C–C–C–C nanoplates

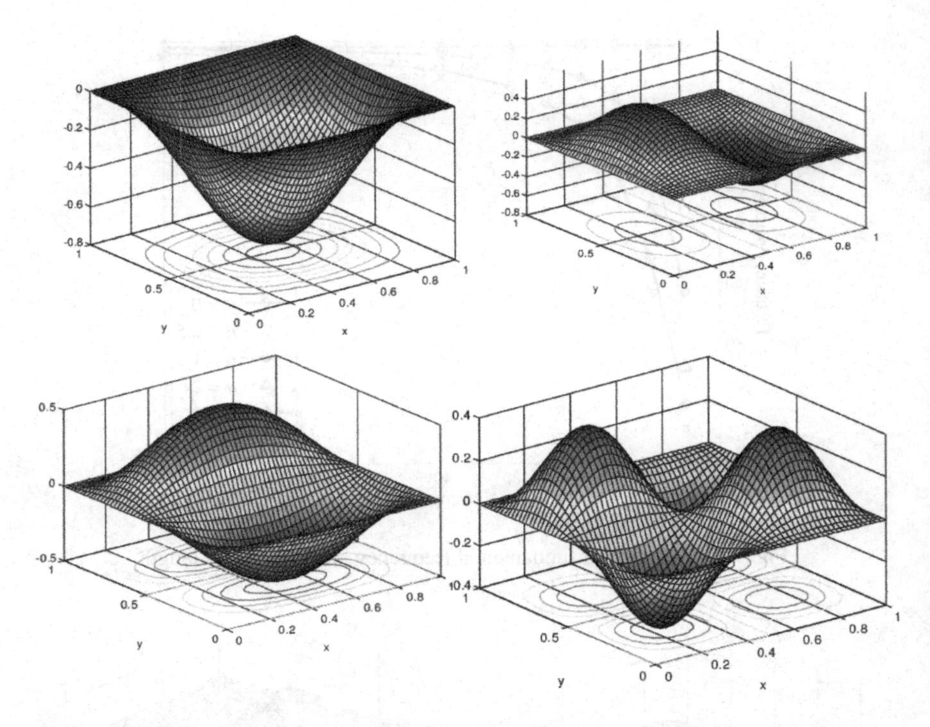

Fig. 9.6 First four deflection shapes of S–C–S–C nanoplates

small scale effect decreases. In this figure, it is also observed that when we increase a, nonlocal curves approach the local curves. This means that for large value of a, all the results converge to those obtained from the classical theory of elasticity. Neglecting nonlocal effect, the relative error percents of fundamental frequency parameter for $a = 10$ nm and $a = 50$ nm with $\mu = 2\ \text{nm}^2$ are 30.61% and 1.91%, respectively. Hence for free vibration of large enough nanoplates, the classical theory of elasticity may be used instead of nonlocal elasticity theory.

9.1.7 *Mode shapes*

Present investigators have reported the first three mode shapes for two sets of boundary conditions such as C–C–C–C, S–C–S–C, and F–S–F–S in Figs. 9.5 and 9.6 with $\mu = 4\ \text{nm}^2$, $a = 5$ nm, and aspect ratio $= 1$.

9.2 Conclusions

Rayleigh–Ritz method is used which can handle any classical boundary conditions at the edges. The present method has been compared with available exact solutions for S–S–S–S boundary condition in graphical form and are found to be in good agreement. Results are given for different nonlocal parameters, length of the nanoplates, and aspect ratios. It is observed that the nonlocality effects should be considered for vibration nanoplates. The present analysis will be of great use to the design engineers who are designing microelectromechanical and nanoelectromechanical devices.

Chapter 10

Vibration of Nanoplates with Complicating Effects

In this chapter, we have shown vibration of nanoplates with complicating effects. Rayleigh–Ritz method with two-dimensional simple polynomials as shape functions have been implemented. Non-uniformity with elastic foundation has been taken into consideration. Parametric studies such as effect of length, aspect ratio, nonlocal parameter, non-uniform parameter, and elastic foundation have been analyzed. Mode shapes for some specified boundary conditions have been presented.

Maximum potential energy may be given as in Eq. (3.35). Maximum kinetic energy may be written as Eq. (3.36).

In this investigation, graphene sheets with non-uniform cross-section have been considered. The non-uniform material properties are assumed as per the following relations:

$$E = E_0(1 + pX + qX^2), \quad \rho = \rho_0(1 + rX + sX^2).$$

Here, we have introduced the non-dimensional variables:

$$X = \frac{x}{a}, \quad Y = \frac{y}{b}, \quad K_w = \frac{k_w a^4}{D}, \quad K_p = \frac{k_p a^2}{D}.$$

Rayleigh quotient in non-dimensional form may be written as

$$\lambda^2 = \frac{\begin{array}{c}\int_0^1\int_0^1 c_a\left[\left(\frac{\partial^2 W}{\partial X^2}\right)^2 + 2\nu R^2\left(\frac{\partial^2 W}{\partial X^2}\frac{\partial^2 W}{\partial Y^2}\right) + R^4\left(\frac{\partial^2 W}{\partial Y^2}\right)^2\right. \\ \left. + 2(1-\nu)R^2\left(\frac{\partial^2 W}{\partial X\partial Y}\right)^2 + c_{kw} + c_{kg}\right] dX\, dY\end{array}}{\int_0^1\int_0^1 c_b\left[W^2 + \frac{\mu}{a^2}\left(\left(\frac{\partial W}{\partial X}\right)^2 + \left(\frac{\partial W}{\partial Y}\right)^2\right)\right] dX\, dY}, \quad (10.1)$$

where

$$\lambda^2 = \frac{\rho_0 h a^4\omega^2}{D_0}, \quad c_a = (1 + pX + qX^2), \quad c_b = (1 + rX + sX^2),$$

$$c_{kw} = K_w\left[W^2 + \frac{\mu}{a^2}\left(\left(\frac{\partial W}{\partial X}\right)^2 + R^2\left(\frac{\partial W}{\partial Y}\right)^2\right)\right],$$

and

$$c_{kg} = K_g\left[\left(\frac{\partial W}{\partial X}\right)^2 + R^2\left(\frac{\partial W}{\partial Y}\right)^2 + \frac{\mu}{a^2}\left(\left(\frac{\partial^2 W}{\partial X^2}\right)^2 + R^2\left(\frac{\partial^2 W}{\partial X\partial Y}\right)^2\right)\right.$$
$$\left. + \frac{\mu}{b^2}\left(\left(\frac{\partial^2 W}{\partial X\partial Y}\right)^2 + R^2\left(\frac{\partial^2 W}{\partial Y^2}\right)^2\right)\right].$$

Using orthonormal polynomials ($\hat{\varphi}_k$) and simple polynomials (φ_k) in Eq. (3.31) and substituting in Eq. (10.1), a generalized eigenvalue problem obtained as

$$[K]\{Z\} = \lambda^2 [M]\{Z\}, \quad (10.2)$$

where Z is a column vector of constants and K and M are the matrices given as follows:

$$K_{ij} = c_a\left[\beta_{ij}^{2020} + \nu R^2\left(\beta_{ij}^{2002} + \beta_{ij}^{0220}\right) + R^4\beta_{ij}^{0202} + 2(1-\nu)R^2\beta_{ij}^{1111}\right.$$
$$\left. + K_w\left[\beta_{ij}^{0000} + \frac{\mu}{a^2}\left(\beta_{ij}^{1010} + R^2\beta_{ij}^{0101}\right)\right]\right]$$
$$+ c_a\left[K_p\left[\beta_{ij}^{1010} + R^2\beta_{ij}^{0101} + \frac{\mu}{a^2}\left(\beta_{ij}^{2020} + R^2\beta_{ij}^{1111}\right)\right.\right.$$
$$\left.\left. + \frac{\mu}{b^2}\left(\beta_{ij}^{1111} + R^2\beta_{ij}^{0202}\right)\right]\right],$$

$$M_{ij} = c_b\left[\beta_{ij}^{0000} + \mu\left(\frac{1}{a}\right)^2\left(\beta_{ij}^{1010} + R^2\beta_{ij}^{0101}\right)\right],$$

with

$$\beta_{ij}^{klmn_a} = \int_0^1 \int_0^1 \left[\frac{\partial^{k+l} \phi_i}{\partial X^k \partial Y^l} \right] \left[\frac{\partial^{m+n_a} \phi_j}{\partial X^m \partial Y^{n_a}} \right] dX\, dY.$$

10.1 Numerical Results and Discussions

Generalized eigenvalue problem (Eq. (10.2)) has been solved and eigenvalues of Eq. (10.1) correspond to the frequency parameters. Different sets of boundary conditions (BCs) have been considered here with Poisson's ratio as 0.3.

10.1.1 *Convergence*

Convergence study of the first three frequency parameters of S–S–S–S and C–C–C–C nanoplates has been shown in Table 10.1 taking $p = 0.2$, $q = 0.3$, $r = 0.4$, $s = 0.5$, aspect ratio $(R) = 1$, nonlocal parameter $(\mu) = 2$ nm^2, and length $(a) = 5$ nm. Results have been shown for $K_w = 0$ and $K_p = 0$. One may see that $n = 37$ is sufficient for computing converged results for nanoplates without elastic foundation. Again, convergence of embedded nanoplates has been shown in Table 10.2 for S–S–S–S edge condition. Numerical values of parameters are taken as $p = q = r = s = 0.1$, $K_w = 200$, $K_p = 5$, $\mu = 1$ nm^2, $a = 10$ nm. It is observed that converged results for embedded nanoplates are obtained

Table 10.1 Convergence of the first three frequency parameters of S–S–S–S and C–C–C–C nanoplates

	S–S–S–S			C–C–C–C		
n	λ_1	λ_2	λ_3	λ_1	λ_2	λ_3
5	11.8785	24.3785	24.4884	19.7881	29.8410	29.9750
10	11.5401	20.9273	24.2786	19.7317	29.0494	29.1283
15	11.5366	20.9013	20.9271	19.7210	29.0469	29.1277
20	11.5366	20.8621	20.8924	19.7210	29.0402	29.1049
25	11.5366	20.8404	20.8620	19.7203	29.0265	29.1048
30	11.5366	20.8404	20.8617	19.7196	29.0264	29.1047
35	11.5366	20.8403	20.8615	19.7196	29.0248	29.1030
36	11.5366	20.8400	20.8615	19.7196	29.0247	29.1030
37	11.5366	20.8400	20.8615	19.7196	29.0247	29.1030

Table 10.2 Convergence of the first three frequency parameters of S–S–S–S nanoplates

n	λ_1	λ_2	λ_3
10	45.6173	63.9934	107.8843
15	45.6044	63.9885	90.5544
20	45.6044	63.8730	90.5418
25	45.6044	63.8729	89.7499
30	45.6044	63.8726	89.7465
35	45.6044	63.8721	89.7464
40	45.6044	63.8721	89.7328
45	45.6044	63.8721	89.7325
46	45.6044	63.8721	89.7325

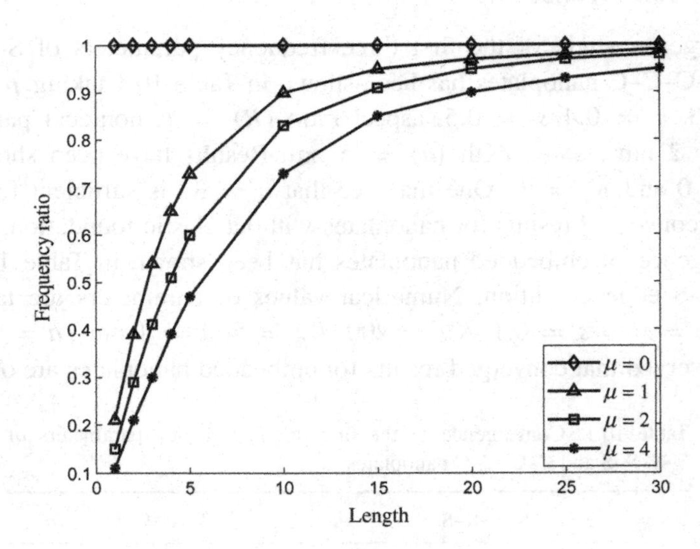

Fig. 10.1 Variation of frequency parameter ratio with length

at $n = 46$. It is also noticed that frequency parameters increase with mode number.

10.1.2 *Validation*

Figures 10.1 and 10.2 show graphical comparisons of S–C–S–C nanoplates with Aksencer and Aydogdu (2011). For this comparison, we have taken $p = q = r = s = K_w = K_p = 0$ with $R = 1$. In these graphs, variation

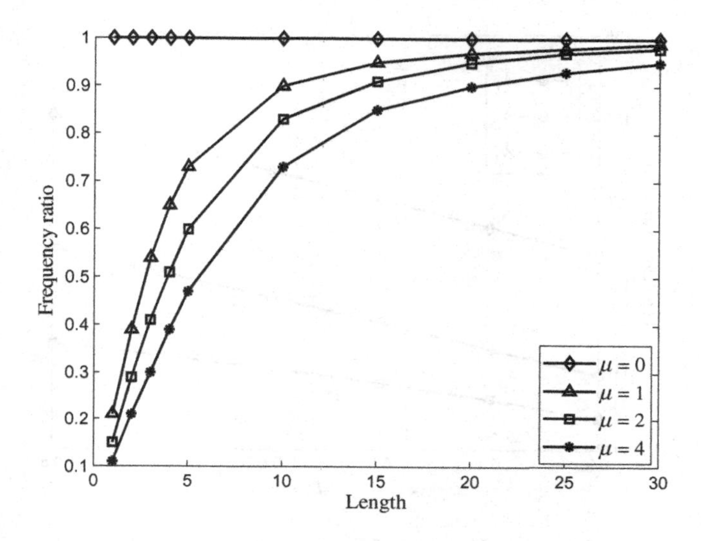

Fig. 10.2 Variation of frequency parameter ratio with length

Table 10.3 Comparison of the first three frequency parameters for S–S–S–S nanoplate

$\mu = 0$		$\mu = 1$		$\mu = 2$		$\mu = 3$		$\mu = 4$	
Present	Ref.*	Present	Ref.*	Present	Ref.*	Present	Ref.*	Present	Ref.*
0.0963	0.0963	0.0880	0.0880	0.0816	0.0816	0.0763	0.0763	0.0720	0.0720
0.3874	0.3853	0.2884	0.288	0.2402	0.2399	0.2102	0.2099	0.1892	0.1889
0.8608	0.8669	0.5167	0.5202	0.4045	0.4063	0.3435	0.3446	0.3037	0.3045

*Aghababaei and Reddy (2009).

of frequency parameter ratio (associated with first two modes) with length has been given for different nonlocal parameters (0, 1, 2, 4 nm^2). Here we have calculated frequency parameter ratio (F_r) as follows:

$$F_r = \frac{\text{frequency parameter calculated using nonlocal theory}}{\text{frequency parameter calculated using local theory}}.$$

One may notice that increase in nonlocal parameter decreases frequency parameter ratio. Same observation may also be seen in Aksencer and Aydogdu (2011) and we may found a close agreement of the results. Next tabular comparison has been given in Table 10.3 with Aghababaei and Reddy (2009) for the first three frequency parameters with $R = 1$

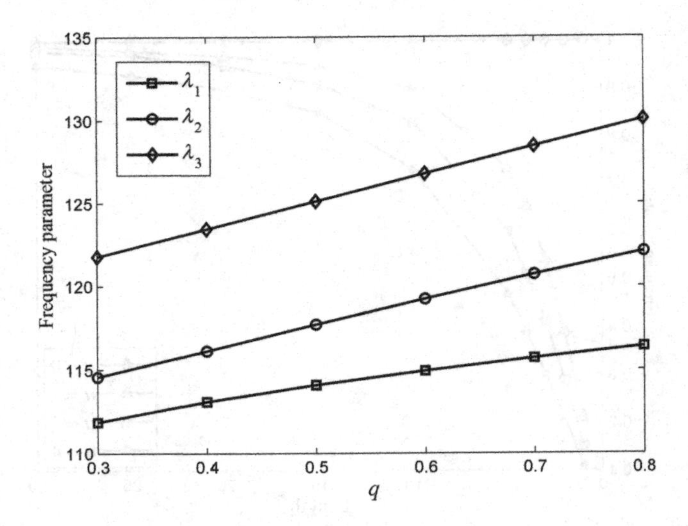

Fig. 10.3 Variation of frequency parameter with q

and $a = 10$ nm. Results have been verified for different nonlocal parameters with the consideration of S–S–S–S edge condition.

10.1.3 *Effect of non-uniform parameter*

In this section, we have studied the effects of non-uniform parameters on the frequency parameters in the absence of elastic foundation. First, we have shown the effects of non-uniform parameters when density and Young's modulus vary quadratically. This case may be achieved by assigning zero to p and r. Variation of the first three frequency parameters with q has been illustrated in Fig. 10.3 keeping s constant (0.3). Similarly, the effect of s on the first three frequency parameters has been shown in Fig. 10.4 keeping q constant (0.2). In these graphs, C–C–C–C edge condition is taken into consideration and the aspect ratio is taken as 2 and 3, respectively. One may see that the frequency parameters decrease with s and increase with q. It is also observed that the frequency parameters increase with increase in mode number.

In this paragraph, we have presented the effects of non-uniform parameters when density and Young's modulus vary linearly. This is achieved by taking q and s to zero. Graphical variation of frequency parameters with p taking r constant (0.3) has been shown in Fig. 10.5. Similarly, graphical

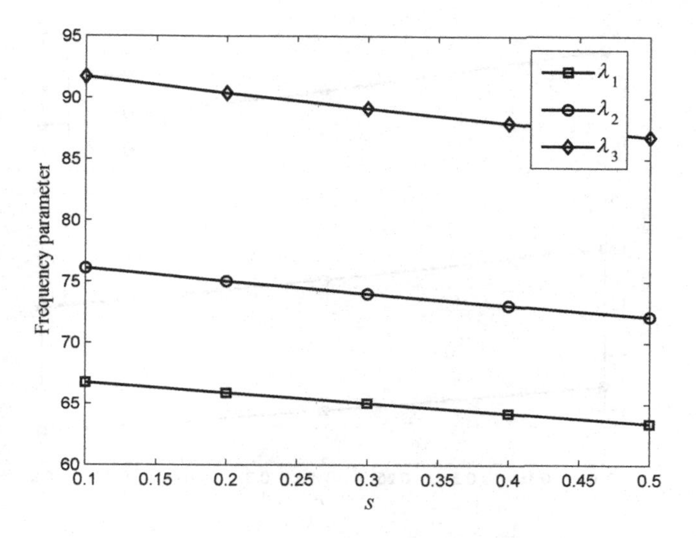

Fig. 10.4 Variation of frequency parameter with s

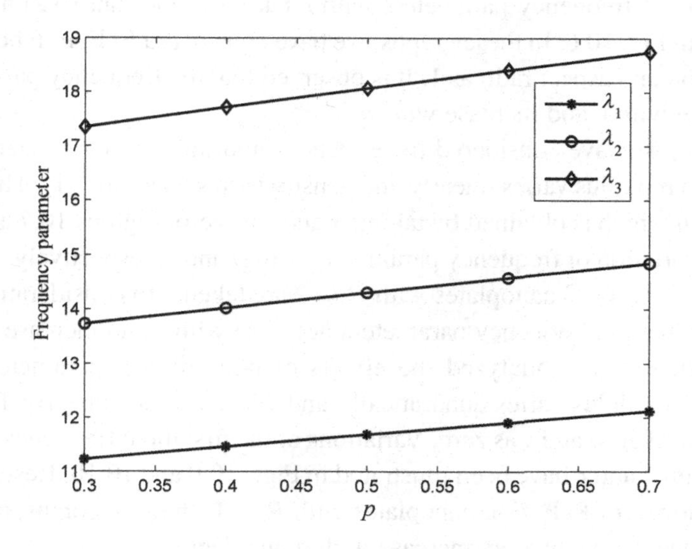

Fig. 10.5 Variation of frequency parameter with p

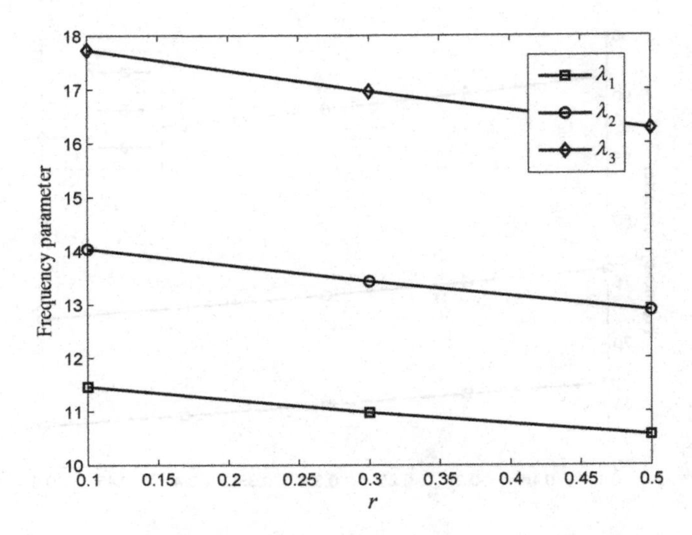

Fig. 10.6 Variation of frequency parameter with r

variation of frequency parameters with r taking p constant (0.2) has been shown in Fig. 10.6. In these graphs, we have considered F–F–F–F boundary condition and aspect ratio as 1. It is observed that the frequency parameters decrease with r and increase with p.

Here, we have considered the effects of non-uniform parameters when Young's modulus varies linearly and density varies quadratically. This is the situation which is obtained by taking q and r as zero. Figures 10.7 and 10.8 depict variation of frequency parameters with p and s, respectively. In these graphs, C–C–C–C nanoplates with $R = 2$ are taken into consideration. It is noticed that the frequency parameters decrease with s and increase with p.

Next, we have analyzed the effects of non-uniform parameter when Young's modulus varies quadratically and density varies linearly. For this, we have taken p and s as zero. Variations of the first three frequency parameters with q and r have been illustrated in Figs. 10.9 and 10.10. Results have been shown for F–F–F–F nanoplates with $R = 1$. In these graphs, one may see frequency parameters increase with q and decrease with r.

10.1.4 *Effect of length*

To investigate the effect of length on the frequency parameters, variation of fundamental frequency parameter with length is shown in Fig. 10.11 for

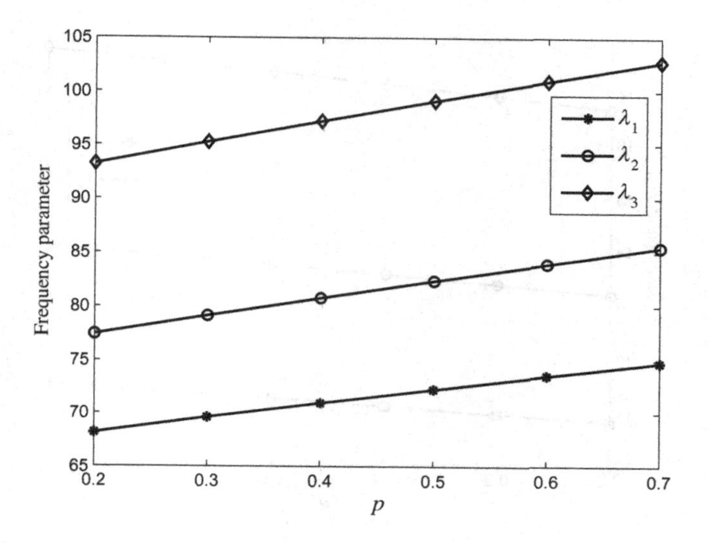

Fig. 10.7 Variation of frequency parameter with p

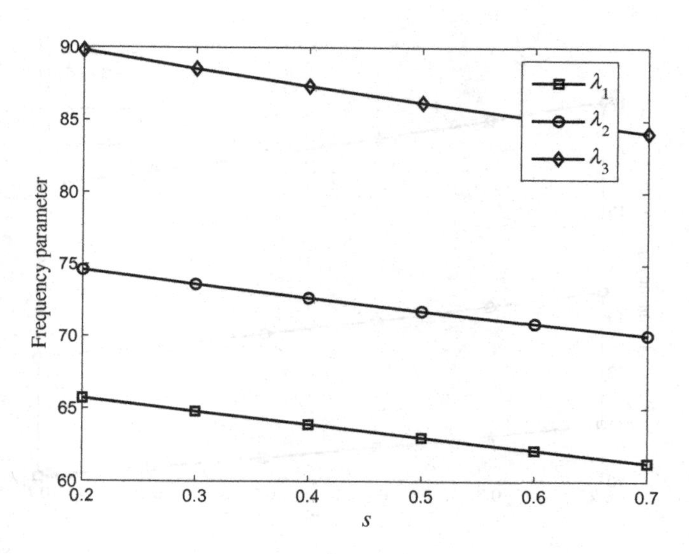

Fig. 10.8 Variation of frequency parameter with s

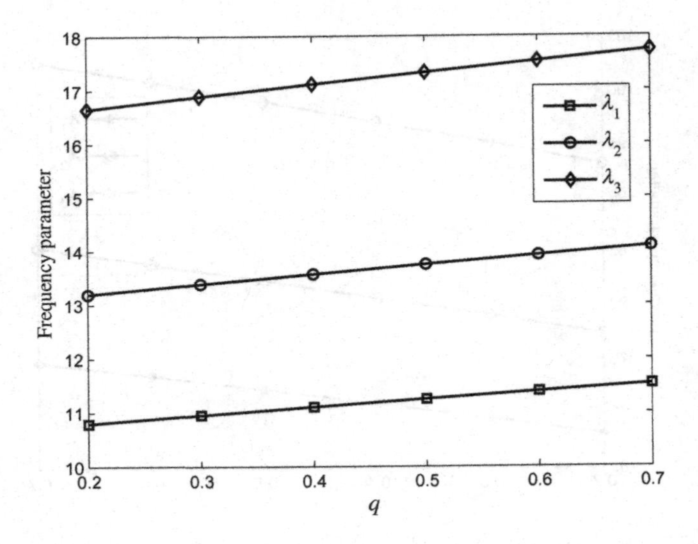

Fig. 10.9 Variation of frequency parameter with q

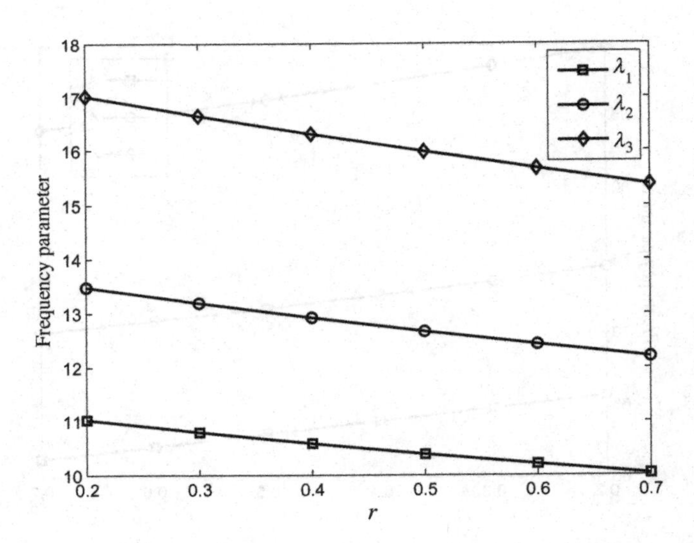

Fig. 10.10 Variation of frequency parameter with r

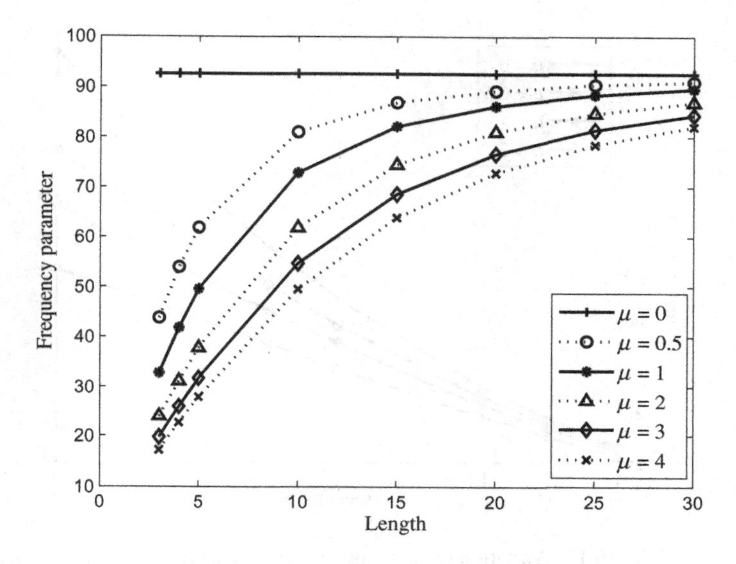

Fig. 10.11 Variation of frequency parameter with length

C–C–C–C nanoplates with $R = 2$, $p = 0.2, q = 0.3, r = 0.4, s = 0.5$ in the absence of elastic foundation. Results have been shown for different values of nonlocal parameters (0, 0.5, 1, 2, 3, 4 nm^2). It is seen that the frequency parameter increases with increase in length. This observation may be explained as follows. Assuming l_{int} as constant, increasing length (a) would lead to decrease in small scale effect (μ/a^2). It is also noticed that the frequency parameters are the highest in the case of $\mu = 0$ and goes on decreasing with increase in nonlocal parameter. This fact may also been explained in terms of relative error percent. Let us define the relative error percent (REP) as

$$\text{REP} = \frac{|\text{ Local result} - \text{nonlocal result }|}{|\text{ Local result }|} \times 100.$$

Neglecting nonlocal effect, REPs of fundamental frequency parameter for $a = 3$ nm and $a = 25$ nm with $\mu = 3$ nm^2 are 78.6753% and 12.0950%, respectively. From this, we may also say that nonlocal theory should be taken into account for vibration analysis of small enough nanoplates.

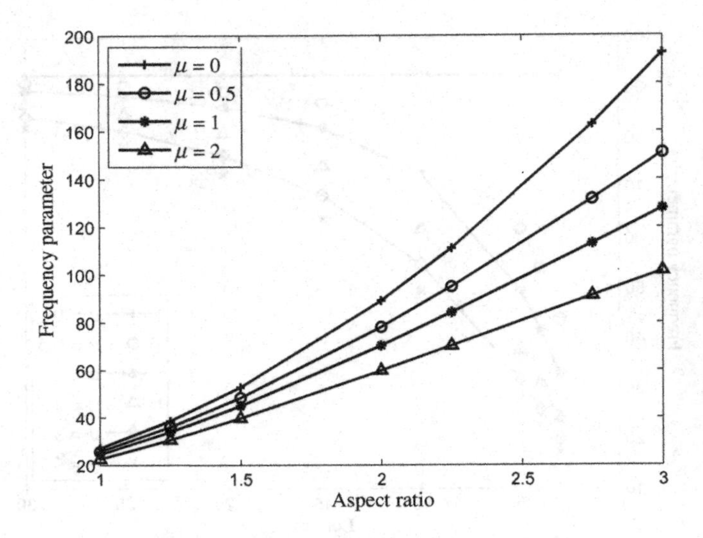

Fig. 10.12 Variation of frequency parameter with aspect ratio

10.1.5 *Effect of aspect ratio*

In this section, we have considered the effect of aspect ratio on the frequency parameters in the absence of elastic foundation. Figure 10.12 shows the effect of fundamental frequency parameter of S–C–S–C nanoplate with aspect ratio taking $a = 10$ nm, $p = 0.1, q = 0.2, r = 0.3, s = 0.4$. It is seen that nonlocal effect on the frequency parameters is more prominent in greater values of aspect ratio. This is due to the fact that for a particular length of nanoplates, increase in aspect ratio would lead to smaller nanoplates which in turn leads to increase in small scale effect. It is also observed that the frequency parameter increases with aspect ratio. One may notice that the frequency parameter decreases with increase in nonlocal parameter. As the nonlocal parameter increases, the frequency parameters obtained from nonlocal plate theory become smaller than those of its local counterpart. This reduction is clearly seen in the case of higher vibration modes. The reduction is due to the fact that nonlocal model may be viewed as atoms linked by elastic springs, while in the case of local continuum model, the spring constant is assumed to take an infinite value. So small scale effect makes the nanoplates more flexible and hence nonlocal impact cannot be neglected. The effect of nonlocal parameter is seen more in the

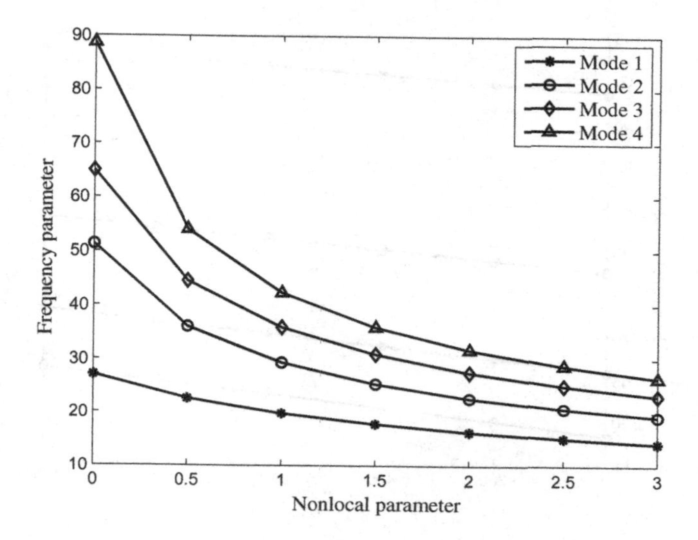

Fig. 10.13 Variation of frequency ratio with nonlocal parameter

case of higher vibration modes. This fact may also be explained in terms of REP. Neglecting nonlocal effect, the REPs for aspect ratios 1 and 3 with $\mu = 3\ \text{nm}^2$ are 22.5041% and 54.9963%, respectively. This shows that nonlocal theory should be considered for free vibration of nanoplates with high aspect ratios.

10.1.6 *Effect of nonlocal parameter*

Here we have examined the effect of nonlocal parameter on the frequency parameters in the absence of elastic foundation. Variation of frequency ratio (associated with the first four modes) with nonlocal parameter has been illustrated in Fig. 10.13 for S–C–S–C edge condition. In this graph, we have taken $a = 5$ nm, $R = 1$, $p = 0.1$, $q = 0.2$, $r = 0.3$, $s = 0.4$. It is clearly seen from the figure that frequency ratio is less than unity. This implies that application of local beam model for vibration analysis of graphene sheets would lead to overprediction of the frequency. Hence, nonlocal beam theory should be used for better predictions of frequencies of nanoplates. Neglecting nonlocal effect, the REPs for the first and fourth modes number of S–C–S–C nanoplates with $\mu = 0.5\ \text{nm}^2$, $R = 1$, $a = 5\,\text{nm}$ are found

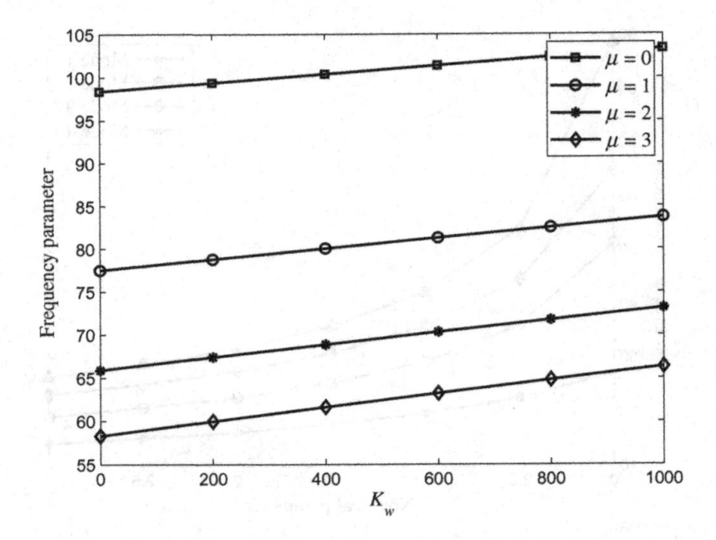

Fig. 10.14 Effect of Winkler coefficient on frequency parameter

to be 15.3605% and 37.8513%, respectively. This shows nonlocal effect on the frequency parameters is more in higher modes.

10.1.7 *Effect of elastic foundation*

In this section, effect of Winkler and Pasternak elastic foundations on the fundamental frequency parameter of embedded nanoplate has been investigated. One may note from Eq. (10.2) that the effects of elastic foundation enter through the stiffness matrix of the nanoplate, i.e. $[K]$. Therefore, the total stiffness of the embedded nanoplate increases as the stiffness of the elastic foundation increases. This trend has been shown in Figs. 10.14 and 10.15 for the springy and shear effect of the elastic foundation, respectively. Numerical values of the parameters are taken as $p = q = r = s = 0.1$, $a = 10$ nm, $R = 2$. In Fig. 10.14, we have considered $K_p = 0$ with C–C–C–C edge condition while in Fig. 10.15, we have taken $K_w = 0$ with S–S–S–S edge condition. Results have been given for different values of nonlocal parameters. It is observed from these figures that the fundamental frequency parameter increases linearly by increasing the stiffness of the elastic foundation either through the springy (Winkler coefficient) or the shear effect (Pasternak coefficient).

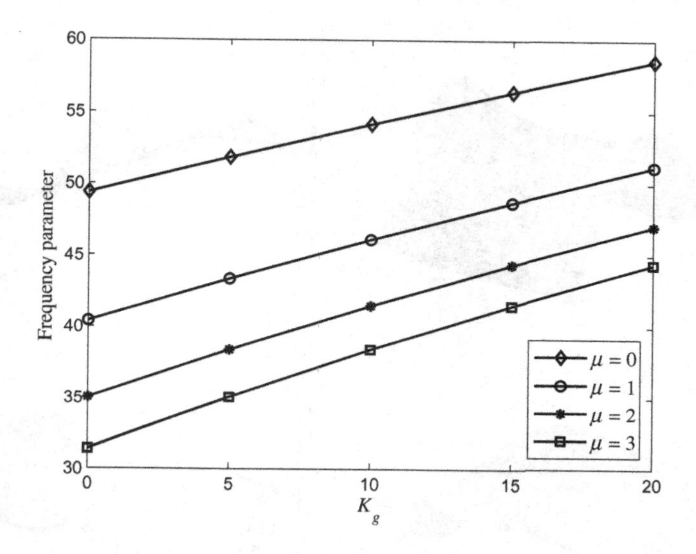

Fig. 10.15 Effect of Pasternak coefficient on frequency parameter

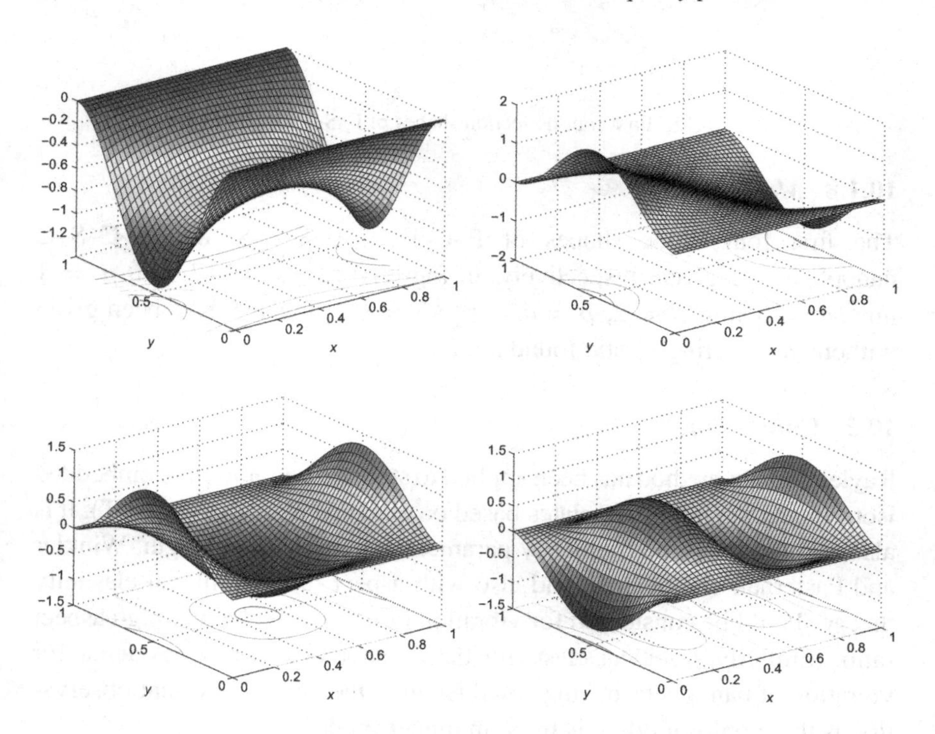

Fig. 10.16 First four deflection shapes of F–C–F–C nanoplates

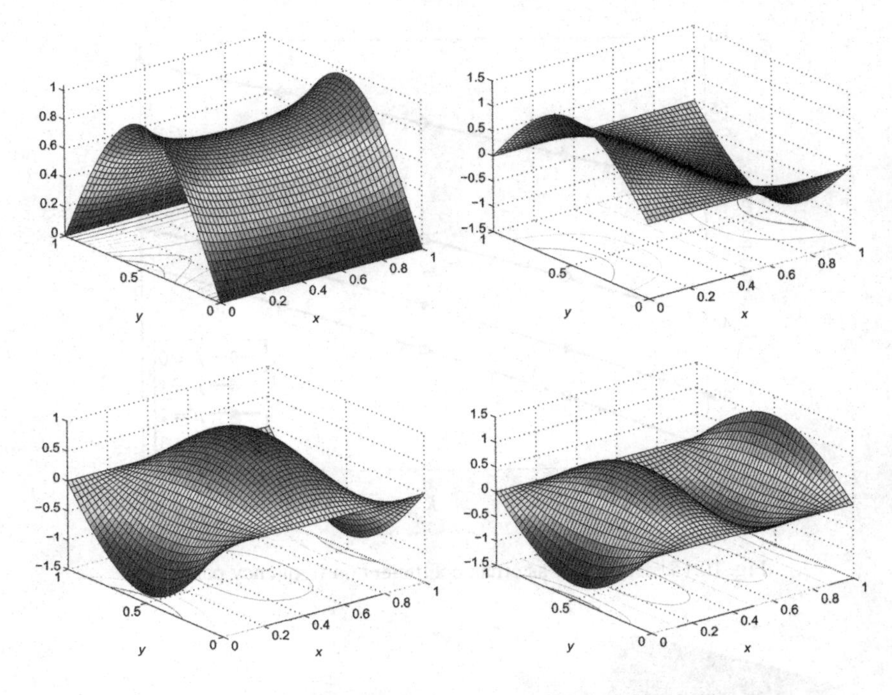

Fig. 10.17 First four deflection shapes of F–S–F–S nanoplates

10.1.8 *Mode shapes*

The first four mode shapes of F–C–F–C, F–S–F–S, and S–C–S–C nanoplates are given, respectively, in Figs. 10.16 and 10.17 with $\mu = 1$ nm^2, $a = 10\,\text{nm}$, $R = 2$, $p = q = r = s = 0.1$. Results have been given without considering elastic foundation.

10.2 Conclusion

Rayleigh–Ritz method has been applied to study free vibration of embedded isotropic rectangular nanoplates based on classical plate theory (CPT). It is also observed that the frequency parameters increase with length, Winkler and Pasternak coefficients, and also with aspect ratio. Nonlocal elasticity theory should be considered for vibration of nanoplates having high aspect ratio. Similarly, nonlocal elasticity theory should also be considered for vibration of nanoplates having small length. One of the important observation is that nonlocal effect is more in higher modes.

Bibliography

Adali, S. (2012) Variational principles for nonlocal continuum model of orthotropic graphene sheets embedded in an elastic medium, *Acta Mathematica Scientia* 32, 325–338.

Aghababaei, R., Reddy, J. N. (2009) Nonlocal third-order shear deformation plate theory with application to bending and vibration of plates, *Journal of Sound and Vibration* 326, 277–289.

Aksencer, T., Aydogdu, M. (2011) Levy type solution method for vibration and buckling of nanoplates using nonlocal elasticity theory, *Physica E* 43, 954–959.

Aksencer, T., Aydogdu, M. (2012) Forced transverse vibration of nanoplates using nonlocal elasticity theory, *Physica E* 44, 1752–1759.

Alibeigloo, A., Zanoosi, A., Pasha, A. (2013) Static analysis of rectangular nano-plate using three-dimensional theory of elasticity, *Applied Mathematical Modelling* 37, 7016–7026.

Alshorbagy, A. E., Eltaher, M. A., Mahmoud, F. F. (2013) Static analysis of nanobeams using nonlocal FEM, *Journal of Mechanical Science and Technology* 27, 2035–2041.

Alzahrani, E. O., Zenkour, A. M., Sobhy, M. (2013) Small scale effect on hygro-thermo-mechanical bending of nanoplates embedded in an elastic medium, *Composite Structures* 105, 163–172.

Amirian, B., Hosseini-Ara, R., Moosavi, H. (2013) Thermal vibration analysis of carbon nanotubes embedded in two-parameter elastic foundation based on nonlocal Timoshenko's beam theory, *Archives of Mechanics* 64, 581–602.

Analooei, H., Azhari, M., Heidarpour, A. (2013) Elastic buckling and vibration analyses of orthotropic nanoplates using nonlocal continuum mechanics and spline finite strip method, *Applied Mathematical Modelling* 37, 6703–6717.

Anjomshoa, A. (2013) Application of Ritz functions in buckling analysis of embedded orthotropic circular and elliptical micro/nano-plates based on nonlocal elasticity theory, *Meccanica* 48, 1337–1353.

Anjomshoa, A., Shahidi, A., Hassani, B., Omehzadeh, E. J. (2014) Finite element buckling analysis of multi-layered graphene sheets on elastic substrate based on nonlocal elasticity theory, *Applied Mathematical Modelling* 38, 5934–5955.

Ansari, R., Gholami, R., Darabi, M. (2011) Thermal buckling analysis of embedded single-walled carbon nanotubes with arbitrary boundary conditions using the nonlocal Timoshenko beam theory, *Journal of Thermal Stresses* 34, 1271–1281.

Ansari, R., Ramezannezhad, H. (2011) Nonlocal Timoshenko beam model for the large-amplitude vibrations of embedded multiwalled carbon nanotubes including thermal effects, *Physica E* 43, 1171–1178.

Ansari, R., Sahmani, S. (2011) Bending behavior and buckling of nanobeams including surface stress effects corresponding to different beam theories, *International Journal of Engineering Science* 49, 1244–1255.

Ansari, R., Sahmani, S. (2012) Small scale effect on vibrational response of single-walled carbon nanotubes with different boundary conditions based on nonlocal beam models, *Communications in Nonlinear Science and Numerical Simulation* 17, 1965–1979.

Ansari, R., Sahmani, S. (2013) Prediction of biaxial buckling behavior of single-layered graphene sheets based on nonlocal plate models and molecular dynamics simulations, *Applied Mathematical Modelling* 37, 7338–7351.

Ansari, R., Sahmani, S., Rouhi, H. (2011a) Axial buckling analysis of single-walled carbon nanotubes in thermal environments via the Rayleigh–Ritz technique, *Computational Materials Science* 50, 3050–3055.

Ansari, R., Sahmani, S., Rouhi, H. (2011b) Rayleigh–Ritz axial buckling analysis of single-walled carbon nanotubes with different boundary conditions, *Physics Letters A* 375, 1255–1263.

Asghari, M., Kahrobaiyan, M. H., Ahmadian, M. T. (2010) A nonlinear Timoshenko beam formulation based on the modified couple stress theory, *International Journal of Engineering Science* 48, 1749–1761.

Aydogdu, M. (2009) A general nonlocal beam theory: Its application to nanobeam bending, buckling and vibration, *Physica E* 41, 1651–1655.

Bedroud, M., Hosseini, H. S., Nazemnezhad, R. (2013) Buckling of circular or annular Mindlin nanoplates via nonlocal elasticity, *Acta Mechanica* 224, 2663–2676.

Behera, L., Chakraverty, S. (2014a) Free vibration of Euler and Timoshenko nanobeams using boundary characteristic orthogonal polynomials, *Applied Nanoscience* 4, 347–358.

Behera, L., Chakraverty, S. (2014b) Free vibration of nonhomogeneous Timoshenko nanobeams, *Meccanica* 49, 51–67.

Benzair, A., Tounsi, A., Besseghier, A., Heireche, H., Moulay, N., Boumia, L. (2008) The thermal effect on vibration of single-walled carbon nanotubes using nonlocal Timoshenko beam theory, *Journal of Physics D: Applied Physics* 41, 225–404.

Bhat, R. B. (1985) Plate deflections using orthogonal polynomials, *Journal of Engineering Mechanics* 111, 1301–1309.

Bhat, R. B. (1991) Vibration of rectangular plates on point and line supports using characteristic orthogonal polynomials in the Rayleigh–Ritz method, *Journal of Sound and Vibration* 149, 170–172.

Chakraverty, S. (2009) *Vibration of Plates*, CRC Press, Taylor & Francis.

Chakraverty, S., Behera, L. (2014) Free vibration of rectangular nanoplates using Rayleigh–Ritz method, *Physica E* 56, 357–363.

Chakraverty, S., Petyt, M. (1997) Natural frequencies for free vibration of nonhomogeneous elliptic and circular plates using two-dimensional orthogonal polynomials, *Applied Mathematical Modelling* 21, 399–417.

Chakraverty, S., Bhat, R. B., Stiharu, I. (1999) Recent research on vibration of structures using boundary characteristic orthogonal polynomials in the Rayleigh–Ritz method, *The Shock and Vibration Digest* 31, 187–194.

Chakraverty, S., Jindal, R., Agarwal, V. K. (2007) Effect of non-homogeneity on natural frequencies of vibration of elliptic plates, *Meccanica* 42, 585–599.

Civalek, Ö., Akgöz, B. (2009) Static analysis of single walled carbon nanotubes based on Eringen's nonlocal elasticity theory, *International Journal of Engineering and Applied Sciences* 1, 47–56.

Civalek, Ö., Demir, C. (2011) Buckling and bending analyses of cantilever carbon nanotubes using the Euler–Bernoulli beam theory based on non-local continuum model, *Asian Journal of Civil Engineering (Building and Housing)* 12, 651–661.

Danesh, M., Farajpour, A., Mohammadi, M. (2012) Axial vibration analysis of a tapered nanorod based on nonlocal elasticity theory and differential quadrature method, *Mechanics Research Communications* 39, 23–27.

Dickinson, S. M. (1978) The buckling and frequency of flexural vibration of rectangular isotropic and orthotropic plates using Rayleigh's method, *Journal of Sound and Vibration* 61, 1–8.

Ehteshami, H., Hajabasi, M. A. (2011) Analytical approaches for vibration analysis of multi-walled carbon nanotubes modeled a multiple nonlocal Euler beams, *Physica E* 44, 270–285.

Eltaher, M. A., Emam, S. A., Mahmoud, F. F. (2012) Free vibration analysis of functionally graded size-dependent nanobeams, *Applied Mathematics and Computation* 218, 7406–7420.

Eltaher, M. A., Emam, S. A., Mahmoud, F. F. (2013a) Static and stability analysis of nonlocal functionally graded nanobeams, *Composite Structures* 96, 82–88.

Eltaher, M. A., Alshorbagy, A. E., Mahmoud F. F. (2013b) Vibration analysis of Euler–Bernoulli nanobeams by using finite element method, *Applied Mathematical Modelling* 37, 4787–4797.

Emam, S. A. (2013) A general nonlocal nonlinear model for buckling of nanobeams, *Applied Mathematical Modelling* 37, 6929–6939.

Eringen, A. C. (1972) Nonlocal polar elastic continua, *International Journal of Engineering Science* 10, 1–16.

Farajpour, A., Danesh, M., Mohammadi, M. (2011) Buckling analysis of variable thickness nanoplates using nonlocal continuum mechanics, *Physica E* 44, 719–727.

Ghannadpour, S. A. M., Mohammadi, B. (2010) Buckling analysis of micro and nano-rods/tubes based on nonlocal Timoshenko beam theory using Chebyshev polynomials, *Advanced Materials Research* 123, 619–622.

Ghannadpour, S. A. M., Mohammadi, B., Fazilati, J. (2013) Bending, buckling and vibration problems of nonlocal Euler beams using Ritz method, *Composite Structures* 96, 584–589.

Hadjesfandiari, A. R., Dargush, G. F. (2011) Couple stress theory for solids, *International Journal of Solids and Structures* 48, 2496–2510.

Hashemi, S. H., Samaei, A. T. (2011) Buckling analysis of micro/nanoscale plates via nonlocal elasticity theory, *Physica E* 43, 1400–1404.

Janghorban, M. (2012) Static analysis of tapered nanowires based on nonlocal Euler–Bernoulli beam theory via differential quadrature method, *Latin American Journal of Solids and Structures* 1, 1–10.

Janghorban, M., Zare, A. (2011) Free vibration analysis of functionally graded carbon nanotubes with variable thickness by differential quadrature method, *Physica E* 43, 1602–1604.

Jomehzadeh, E., Saidi, A. R. (2012) Study of small scale effect on nonlinear vibration of nano-plates, *Journal of Computational and Theoretical Nanoscience* 9, 864–871.

Kananipour, H. (2014) Static analysis of nanoplates based on the nonlocal Kirchhoff and Mindlin plate theories using DQM, *Latin American Journal of Solids and Structures* 11, 1709–1720.

Ke, L., Xiang, Y., Yang, J., Kitipornchai, S. (2009) Nonlinear free vibration of embedded double-walled carbon nanotubes based on nonlocal Timoshenko beam theory, *Computational Materials Science* 47, 409–417.

Kiani, K. (2011) Small-scale effect on the vibration of thin nanoplates subjected to a moving nanoparticle via nonlocal continuum theory, *Journal of Sound and Vibration* 330, 4896–4914.

Kumar, D., Heinrich, C., Waas, A. M. (2008) Buckling analysis of carbon nanotubes modeled using nonlocal continuum theories, *Journal of Applied Physics* 103, 073521.

Lee, H. L., Chang W. J. (2009) A closed-form solution for critical buckling temperature of a single-walled carbon nanotube, *Physica E* 41, 1492–1494.

Liu, C., Ke, L., Wang, Y. S., Yang, J., Kitipornchai, S. (2013) Thermo-electro-mechanical vibration of piezoelectric nanoplates based on the nonlocal theory, *Composite Structures* 106, 167–174.

Loya, J., López-Puente, J., Zaera, R., Fernández-Sáez, J. (2009) Free transverse vibrations of cracked nanobeams using a nonlocal elasticity model, *Journal of Applied Physics* 105, 044309.

Lu, P., Lee, H., Lu, C., Zhang, P. (2006) Dynamic properties of flexural beams using a nonlocal elasticity model, *Journal of Applied Physics* 99, 073510.

Lu, P., Lee, H. P., Lu, C., Zhang, P. Q. (2007) Application of nonlocal beam models for carbon nanotubes, *International Journal of Solids and Structures* 44, 5289–5300.

Maachou, M., Zidour, M., Baghdadi, H., Ziane, N., Tounsi, A. (2011) A nonlocal Levinson beam model for free vibration analysis of zigzag single-walled carbon nanotubes including thermal effects, *Solid State Communications* 151, 1467–1471.

Mahmoud, F. F., Eltaher, M. A., Alshorbagy, A. E., Meletis, E. I. (2012) Static analysis of nanobeams including surface effects by nonlocal finite element, *Journal of Mechanical Science and Technology* 26, 3555–3563.

Malekzadeh, P., Shojaee, M. (2013) Free vibration of nanoplates based on a nonlocal two-variable refined plate theory, *Composite Structures* 95, 443–452.

Malekzadeh, P., Setoodeh, A. R., Beni, A. A. (2011a) Small scale effect on the free vibration of orthotropic arbitrary straight-sided quadrilateral nanoplates, *Composite Structures* 93, 1631–1639.

Malekzadeh, P., Setoodeh, A. R., Beni, A. (2011b) Small scale effect on the thermal buckling of orthotropic arbitrary straight-sided quadrilateral nanoplates embedded in an elastic medium, *Composite Structures* 93, 2083–2089.

Mohammadi, B., Ghannadpour, S. (2011) Energy approach vibration analysis of nonlocal Timoshenko beam theory, *Procedia Engineering* 10, 1766–1771.

Murmu, T., Adhikari, S. (2010a) Nonlocal transverse vibration of double-nanobeam-systems, *Journal of Applied Physics* 108, 083514.

Murmu, T., Adhikari, S. (2010b) Scale-dependent vibration analysis of prestressed carbon nanotubes undergoing rotation, *Journal of Applied Physics* 8, 123507.

Murmu, T., Adhikari, S. (2009) Small-scale effect on the vibration of nonuniform nanocantilever based on nonlocal elasticity theory, *Physica E* 41, 1451–1456.

Murmu, T., Pradhan, S. C. (2009a) Thermo-mechanical vibration of a single-walled carbon nanotube embedded in an elastic medium based on nonlocal elasticity theory, *Computational Materials Science* 46, 854–859.

Murmu, T., Pradhan, S. C. (2009b) Buckling analysis of a single-walled carbon nanotube embedded in an elastic medium based on nonlocal elasticity and Timoshenko beam theory and using DQM, *Physica E* 41, 1232–1239.

Murmu, T., Pradhan, S. C. (2009c) Vibration analysis of nanoplates under uniaxial prestressed conditions via nonlocal elasticity, *Journal of Applied Physics* 106, 104301.

Murmu, T., Pradhan, S. C. (2009d) Vibration analysis of nano-single layered graphene sheets embedded in elastic medium based on nonlocal elasticity theory, *Journal of Applied Physics* 105, 06431.

Murmu, T., Pradhan, S. C. (2010) Thermal effects on the stability of embedded carbon nanotubes, *Computational Materials Science* 47, 721–726.

Mustapha, K. B., Zhong, Z. W. (2010) Free transverse vibration of an axially loaded non-prismatic single-walled carbon nanotube embedded in a two-parameter elastic medium, *Computational Materials Science* 50, 742–751.

Naderi, A., Saidi, A. R. (2014) Modified nonlocal Mindlin plate theory for buckling analysis of nanoplates, *Journal of Nanomechanics and Micromechanics* 4, 2153–5477.

Nami, M. R., Janghorban, M. (2013) Static analysis of rectangular nanoplates using trigonometric shear deformation theory based on nonlocal elasticity theory, *Beilstein Journal of Nanotechnology* 4, 968–973.

Nami, M. R., Janghorban M. (2014) Static analysis of rectangular nanoplates using exponential shear deformation theory based on strain gradient elasticity theory, *Iranian Journal of Materials Forming* 1, 1–13.

Narendar, S. (2011) Buckling analysis of micro-/nano-scale plates based on two-variable refined plate theory incorporating nonlocal scale effects, *Composite Structures* 93, 3093–3103.

Narendar, S., Gopalakrishnan, S. (2011) Critical buckling temperature of single-walled carbon nanotubes embedded in a one-parameter elastic medium based on nonlocal continuum mechanics, *Physica E* 43, 1185–1191.

Narendar, S., Gopalakrishnan, S. (2012) Scale effects on buckling analysis of orthotropic nanoplates based on nonlocal two-variable refined plate theory, *Acta Mechanica* 223, 395–413.

Nix, W. D., Gao, H. (1998) Indentation size effects in crystalline materials: A law for strain gradient plasticity, *Journal of the Mechanics and Physics of Solids* 46, 411–425.

Peddieson, J., Buchanan, G. R., McNitt, R. P. (2003) Application of nonlocal continuum models to nanotechnology, *International Journal of Engineering Science* 41, 305–312.

Phadikar, J. K., Pradhan, S. C. (2010) Variational formulation and finite element analysis for nonlocal elastic nanobeams and nanoplates, *Computational Materials Science* 49, 492–499.

Pradhan, S. C. (2009) Buckling of single layer graphene sheet based on nonlocal elasticity and higher order shear deformation theory, *Physics Letters A* 373, 4182–4188.

Pradhan, S. C. (2012) Buckling analysis and small scale effect of biaxially compressed graphene sheets using non-local elasticity theory, *Sadhana* 37, 461–480.

Pradhan S. C., Murmu T. (2010a) Small scale effect on the buckling analysis of single-layered graphene sheet embedded in an elastic medium based on nonlocal plate theory, *Physica E* 42, 1293–1301.

Pradhan, S. C., Murmu, T. (2010b) Application of nonlocal elasticity and DQM in the flapwise bending vibration of a rotating nanocantilever, *Physica E* 42, 1944–1949.

Pradhan, S. C., Phadikar, J. K. (2009a) Bending, buckling and vibration analyses of nonhomogeneous nanotubes using GDQ and nonlocal elasticity theory, *Structural Engineering and Mechanics* 33, 193–213.

Pradhan, S. C., Phadikar, J. K. (2009b) Nonlocal elasticity theory for vibration of nanoplates, *Journal of Sound and Vibration* 325, 206–223.

Pradhan, S. C., Reddy, G. (2011) Buckling analysis of single walled carbon nanotube on Winkler foundation using nonlocal elasticity theory and DTM, *Computational Materials Science* 50, 1052–1056.

Quan, J, Chang, C. (1989) New insights in solving distributed system equations by the quadrature method-I, *Computers and Chemical Engineering* 13, 779–788.

Rafiei, M., Mohebpour, S. R., Daneshmand, F. (2012) Small-scale effect on the vibration of non-uniform carbon nanotubes conveying fluid and embedded in viscoelastic medium, *Physica E* 44, 1372–1379.

Ravari, M. R. K., Shahidi, A. R. (2013) Axisymmetric buckling of the circular annular nanoplates using finite difference method, *Meccanica* 48, 135–144.

Reddy, J. N. (1997) *Mechanics of Laminated Composite Plates: Theory and Analysis*, CRC Press, Boca Raton, FL.

Reddy, J. N. (2007) Nonlocal theories for bending, buckling and vibration of beams, *International Journal of Engineering Science* 45, 288–307.

Reddy, J. N., Pang, S. D. (2008) Nonlocal continuum theories of beams for the analysis of carbon nanotubes, *Journal of Applied Physics* 103, 023511.

Roque, C., Ferreira, A., Reddy, J. (2011) Analysis of Timoshenko nanobeams with a nonlocal formulation and meshless method, *International Journal of Engineering Science* 49, 976–984.

Sahmani, S., Ansari, R. (2011) Nonlocal beam models for buckling of nanobeams using state-space method regarding different boundary conditions, *Journal of Mechanical Science and Technology* 25, 2365–2375.

Setoodeh, A. R., Malekzadeh, P., Vosoughi, A. R. (2011) Nonlinear free vibration of orthotropic graphene sheets using nonlocal Mindlin plate theory, *Proceedings of the Institution of Mechanical Engineers, Part C: J Mechanical Engineering Science* 226, 1896–1906.

Şimşek, M., Yurtcu, H. H. (2013) Analytical solutions for bending and buckling of functionally graded nanobeams based on the nonlocal Timoshenko beam theory, *Composite Structures* 97, 378–386.

Singh, B., Chakraverty, S. (1992) Transverse vibration of simply supported elliptical and circular plates using boundary characteristic orthogonal polynomials in two variables, *Journal of Sound and Vibration* 152, 149–155.

Singh, B., Chakraverty, S. (1994a) Boundary characteristic orthogonal polynomials in numerical approximation, *Communications in Numerical Methods in Engineering* 10, 1027–1043.

Singh, B., Chakraverty, S. (1994b) Flexural vibration of skew plates using characteristic orthogonal polynomials in two variables, *Journal of Sound and Vibration* 173, 157–178.

Thai, H. T. (2012) A nonlocal beam theory for bending, buckling, and vibration of nanobeams, *International Journal of Engineering Science* 52, 56–64.

Thai, H. T., Vo, T. P. (2012) A nonlocal sinusoidal shear deformation beam theory with application to bending, buckling, and vibration of nanobeams, *International Journal of Engineering Science* 54, 58–66.

Tounsi, A., Semmah, A., Bousahla, A. A. (2013) Thermal buckling behavior of nanobeams using an efficient higher-order nonlocal beam theory, *Journal of Nanomechanics and Micromechanics* 3, 37–42.

Wang, X., Bert, C. W. (1993) A new approach in applying differential quadrature to static and free vibration analyses of beam and plates, *Journal of Sound and Vibration* 162, 566–572.

Wang, Q., Varadan, V. K. (2006) Vibration of carbon nanotubes studied using nonlocal continuum mechanics, *Smart Material Structures* 15, 659–666.

Wang, K. F., Wang, B. L. (2011) Vibration of nanoscale plates with surface energy via nonlocal elasticity, *Physica E* 44, 448–453.

Wang, C. M., Reddy, J. N., Lee, K. H. (2000) *Shear Deformable Beams and Plates: Relationship with Classical Solutions*, Elsevier.

Wang, C. M., Zhang, Y. Y., He, X. Q. (2007) Vibration of nonlocal Timoshenko beams, *Nanotechnology* 18, 105401.

Wang, C. M., Kitipornchai, S., Lim, C. W., Eisenberger, M. (2008a) Beam bending solutions based on nonlocal Timoshenko beam theory, *Journal of Engineering Mechanics* 134, 475–481.

Wang, L., Ni, Q., Li, M., Qian Q. (2008b) The thermal effect on vibration and instability of carbon nanotubes conveying fluid, *Physica E* 40, 3179–3182.

Wang, C. M., Zhang, Y. Y., Ramesh, S. S., Kitipornchai, S. (2006) Buckling analysis of micro-and nano-rods/tubes based on nonlocal Timoshenko beam theory, *Journal of Physics D: Applied Physics* 39, 3904.

Wang, C. Y., Murmu, T., Adhikari, S. (2011) Mechanisms of nonlocal effect on the vibration of nanoplates, *Applied Physics Letters* 98, 153101.

Xu, M. (2006) Free transverse vibrations of nano-to-micron scale, *Proceedings of the Royal Society A* 462, 2977–2995.

Yan, Y., Wang, W., Zhang, L. (2010) Nonlocal effect on axially compressed buckling of triple-walled carbon nanotubes under temperature field, *Applied Mathematical Modelling* 34, 3422–3429.

Yang, J., Ke, L., Kitipornchai, S. (2010) Nonlinear free vibration of single-walled carbon nanotubes using nonlocal Timoshenko beam theory, *Physica E* 42, 1727–1735.

Yang, Y., Lim, C. (2011) A variational principle approach for buckling of carbon nanotubes based on nonlocal Timoshenko beam models, *Nano* 6, 363–377.

Zenkour, A. M., Sobhy, M. (2013) Nonlocal elasticity theory for thermal buckling of nanoplates lying on Winkler–Pasternak elastic substrate medium, *Physica E* 53, 251–259.

Zhang, Y., Liu, G., Xie, X. (2005) Free transverse vibrations of double-walled carbon nanotubes using a theory of nonlocal elasticity, *Physical Review B* 71, 195404.

Zhang, Y., Liu, X., Liu, G. (2007) Thermal effect on transverse vibrations of double-walled carbon nanotubes, *Nanotechnology* 18, 445701.

Zidour, M., Benrahou, K. H., Semmah, A., Naceri, M., Belhadj, H. A., Bakhti, K., Tounsi, A. (2012) The thermal effect on vibration of zigzag single walled carbon nanotubes using nonlocal Timoshenko beam theory, *Computational Materials Science* 51, 252–260.

Index

O

P

R

S

T

V

W

Y